Recent Research in Psychology

Arno L. Goudsmit (Ed.)

Self-Organization in Psychotherapy

Demarcations of a New Perspective

Springer-Verlag
Berlin Heidelberg New York
London Paris Tokyo Hong Kong

Editor

Arno L. Goudsmit
Dept. of Foundations and History of Psychology
University of Groningen
P. O. Box 72, 9700 AB Groningen
The Netherlands

ISBN 978-3-540-52161-7 ISBN 978-3-642-48704-0 (eBook)
DOI 10.1007/978-3-642-48704-0

2126/3140-543210 – Printed on acid-free paper

Contents

1

By Way of Introduction: Can We Make a Non-Classical Psychology?

Arno L. Goudsmit

THERE IS A basic split between two traditions in psychology. We may call them the 'outside tradition' and the 'inside tradition'. The following terms connote these traditions:

'outside tradition':	*'inside tradition'*:
positivistic	phenomenological
reductionistic	holistic
explaining	understanding
experimental	descriptive
nomothetic	idiographic
-etic	-emic
quantitative	qualitative
objective	subjective
'hard'	'soft'

These two research traditions, based on much older philosophical traditions, have developed during this century into more and more mutually opposing movements in psychology, each having its own strong and weak sides. Usually in debates between the two movements the strong sides of one are contrasted to the weak sides of the other. Also some 'reconciliating' styles of research have been developed, in which it is tried to combine the strong sides of both, e.g. by having a large quantitative survey study be preceded by a small qualitative exploratory 'pilot' study. No attempts seem to have been made to relate the weak sides of both traditions. Why should one?

The aim of this introduction is to do just this: to relate the weak sides of both, in order to show that these weaknesses can be seen to converge precisely toward one central issue, that usually evades attention. This central issue will be described in terms of a particular relation between object and method, one that plays a central role in understanding psychotherapeutic processes in terms of self-organization.

A Metaphor from Quantum Physics and from Phenomenology

It usually goes unnoticed that two so remote authors as Niels Bohr and Maurice Merleau-Ponty both make use of a particular metaphor, in discussing aspects of human perception. Bohr (1934, p. 99), being concerned with the boundaries between the observer and the physical object he is investigating, had a need to express the impossibility of observing one's measuring instrument while using it as such. This was particularly relevant in his description of the role played by the measuring device while observing quantum phenomena. Merleau-Ponty (1945, p. 167) was also concerned with the boundary between a sensing individual and his environment, with regard to how an individual organizes his environment in respect of his own body. Both authors use, in order to make their point, a metaphor which is quite interesting for our present purposes. It is the metaphor of a stick that can be used as an instrument of touch.

Bohr speaks of a stick that, if grasped firmly, can be used as an instrument for touching objects in the environment. The tactile sensations in the hand then escape from attention, and instead the distal edge of the stick takes the quality of a tactile organ. It is there, at this distal site, where the person observes the object he is touching with his stick; no longer is the palm of the hand the boundary of the person as a sensing unity, but the edge of the stick. Conversely, if the stick is held loosely, it cannot be used as an instrument of touch, and it appears to the observer as a stick, i.e.: as an independent object, sensed in the hand. Likewise, Merleau-Ponty gives the example of a blind man using a stick as an elongation of his own body. Thus, the blind man has his stick 'participate' in his own body. Then the way in which he perceives his environment by means of the stick necessarily evades his attention. The stick has become incorporated.

Both for phenomenologists and for quantum theorists the metaphor illustrates that we cannot simultaneously pay attention both to an object and to the method by which it is perceived. Bohr is considered to have been inspired in this respect by William James (cf. Holton, 1973).

According to this basic complementarity between method and object of perception, there seem to be two possible courses of action: either to focus upon the object, and allow the instrument to become part of one's 'tacit knowledge' (Polanyi, e.g. 1966), or to keep one's mind clear about the method, irrespective of the objects that one will face. This is a matter of *priority*.

It is the (implicit) assignment of this priority which determines in which tradition we operate. Priority of object over method has been assumed by the 'inside' tradition, and the reverse priority has been assumed by the 'outside' tradition. It is after having made this assignment of priority, that substantial problems arise for those wanting to overcome the object-method complementarity. Both Merleau-Ponty and Bohr, as representatives from either tradition, explored the world as it is before this priority assignment has been made. This is what qualifies their works as *non-classical*. It is this kind of non-classicality that, in my opinion, is needed in psychology. Below I will give a small crumb of a sketch for such a non-classical psychology. I will describe the object-method complementarity in terms of a relation between the priority assignment and the weaknesses of both the 'inside' and the 'outside' tradition. First, let me give a look at these weaknesses.

The Weakness of the 'Inside' Tradition: the Method Fades Out

The inside tradition has been defined in various ways. Shotter's description of 'practical knowledge' is the most proper to our present purposes. He describes it as:

> "... *knowing from within a situation, which takes into account, in what is known, the situation within which it is known.*" (Shotter, 1985, p. 448).

Adherents of the inside tradition may or may not differ in respect of whether the object they are interested in exists independently of their knowing acts. What matters is that the object is given priority to the method. Here we encounter a problem. The observer is considered to influence, if not to constitute, his object of study by his knowing acts. However, a specification by the observer of the context in which the object exists would also comprise his act of knowing the object.

The inside tradition has thus as a logical limit the observer's attempt to specify fully the method itself as a 'contextual aspect' of the object of study. This is why Polanyi's term 'tacit knowledge' is felicitous: the observer will never end in making the context explicit. His inside knowledge of the object seems to consist of an infinite hierarchy of implicit abilities, each of which is concerned with putting into practice the knowledge that has already been made explicit.

Examples:

a.
A practicing psychotherapist tries to be explicit about what he is doing when he is empathic with a patient. As soon as he starts to regard his empathy as a technique that can be performed, he finds himself in need of pointing at his own private experiences, from which he was able to utilize this 'technique' appropriately. The crux

of empathy, and of authenticity in general, seems to escape whenever one tries to formulate it as an executable technique.

b.
A participant observer tries to be explicit about his performance, e.g. in an anthropological field study. Since his behavior is part of the situation studied (e.g. a party), description of his research behavior is necessarily embedded in a description of the situation. The method is absorbed by the object of study. Extricating it from the object can only be done at the cost of no longer understanding why this 'procedure' has been followed.

The weakness of this tradition, thus, resides in the observer's incapacity to account fully for the method by which he obtained his 'inside knowledge'. Rather, he feels compelled to declare his knowledge to be of too much a 'contextual' nature as to formulate it as a fully explicit and reproducible procedure. The method is absorbed by the object of study, and cannot be extricated from the object. The object has priority over the method. The 'pure' method can only be formulated in abstract terms, as an in itself impracticable procedure. Wouldn't the availability of a clear and context-free method free us from many impediments?

The Weakness of the 'Outside' Tradition: the Object Fades Out

The 'outside' tradition is the mainstream scientific culture. It assigns a predominant role to the development and elaboration of scientific methods. Only by being critical about the way one conducts one's observations and argumentations, it is thought possible for an observer to arrive at knowledge that is of a justifiable degree of certainty. If the method cannot be accounted for, then the insights thus obtained are considered gratuitous.

Adherents of the outside tradition may or may not differ in respect of whether the object they are interested in exists independently from their acts of measuring it. What matters is that the method is given priority to the object. It is thought virtually possible to describe the world by means of an axiomatized system ('more geometrico'), i.e., by means of a method we are fully aware of, as we apply it. The method can also be conceived apart from that to which it is applied.

Since the method is kept here continuously under critical control, the weakness of this tradition does not reside, as above, in a lack of explicit procedural knowledge. To the contrary, the research techniques are clear and accessible. This time, however, the problem is in the object of investigation.
Many aspects of the 'outside' techniques and methods, therefore, are aimed at establishing a clear image of the object, distinct from the method itself. The bulk of statistical techniques for example is designed to do just this. It allows the researcher to 'subtract' the assessed features of the measuring procedure from the 'raw data', so that what remains may validly be interpreted as features of the object of investigation. This kind of 'subtraction thought' is beautifully illustrated by the language used in the following fragment:

> "*If, after determining that neither concept redefinition nor scale recalibration has occurred, a researcher observes a difference in subject responses from time-1 to time-2, behavioral change can be said to have been detected.*" (Armenakis, 1988, p. 165)

For adherents of this tradition it is of the foremost importance that the object can be distinguished from the measuring instrument. Statistical criteria are often available to calculate measurement errors, and to decide whether or not measurements validly and reliably represent features of the object of study.

The outside tradition has thus as a logical limit the observer's attempt to specify fully the object itself as the outcome of his method. Then the observer becomes more and more entangled in technical issues of analysis, instead of being able to perceive his object by means of his method. The 'error terms' become too high, the 'signal-noise ratio' too low, or the sample too small. As a result, the focal object fades out.

The less the object can be distinguished from the method, the more the latter is given priority over the former[1]. In a sense, then, the object is absorbed by the method and can only be formulated as an abstract concept.

Object-Method Complementarity and the Relation between both Traditions

The validity and reliability issue is for the outside tradition the Achilles heel, as is the context issue for the inside tradition. The weakness of the inside tradition resided in that the method was absorbed by the object; the object was no longer a background against which the method could be delineated. Likewise, the weakness of the outside tradition resided in that the object was absorbed by the method.

What the weaknesses of both traditions have in common is their convergence to indistinguishability[2] between object and method. We say that the method cannot be distinguished from the object, if the object absorbs the method. In this way we are able to formulate the weaknesses of both traditions in terms of the amount of difficulties it takes the observer to distinguish object and method from one another.

We may put our two traditions thus on a horizontal 'distinguishability' dimension (see figure 1), linking the two traditions together at their weakest spots. Here 'weak inside' means that many difficulties arise for the observer in distinguishing method from object. 'Weak outside' means the same for the reverse distinction: object from method.

The more difficulties arise in making either distinction, the more the 'priority' that is assumed by a tradition, becomes relevant. This 'priority' is given by the observer in accordance with the tradition he

---

[1]Another example of priority of method over object, can be found in Kelley et al. (1983), as criticized by Shotter (1987).

[2]Notice that such a distinction means that object and method are not distinguished as entities 'in themselves', but always as a figure against a background. Either may take the role of figure or background.

is in. 'Priority of object over method', for example, is the degree to which the observer attempts to keep sight of his object of investigation. The priority value, therefore, denotes that to which the observer clasps as he looses the ability to make distinctions.

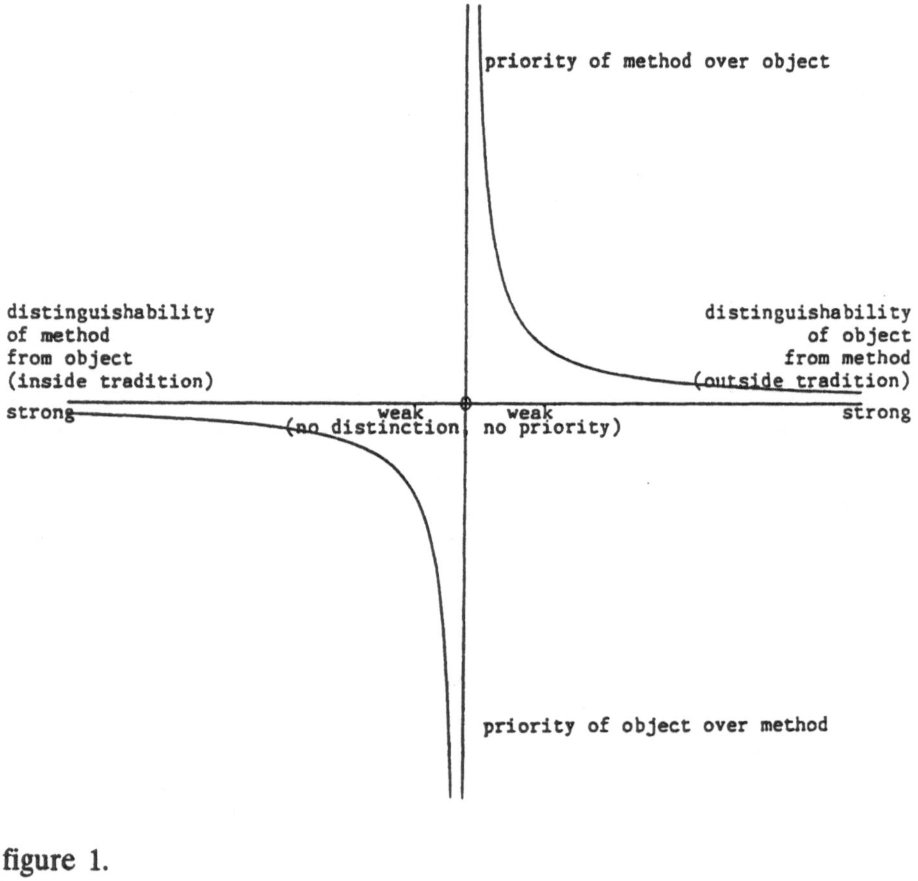

figure 1.

Figure 1 shows 'priority' as a function of 'distinguishability'. Then at the 'weakest' value of both traditions the 'priority' values (i.e.: the priority of object over method, as well as the priority of method over object) asymptotically tend to infinite values. That is, at the value of 'no distinction' between object and method, 'priority' is undefined. For the present purposes of conceptual definition, the relation between 'distinguishability' and 'priority' may be sketched as a hyperbola of the type: $y = 1/x$. Clearly, at extremely 'strong' values of 'distinguishability', the 'priority' values asymptotically tend to the 'no priority' value: Object and method are clearly distinguishable.

The Domain of a Non-Classical Psychology

Notice that both dimensions 'distinguishability' and 'priority' are defined in relation to the observer who attempts to distinguish method and object. They are not defined as ontological 'an sich' qualities of method-object relations, that would exist irrespective of an observer.

Classical psychology[3] in fact does make such an ontological assumption, viz. that object and method belong to distinct categories: both are conceived to exist as observer-independent entities (persons, things, or tools, programs, etc.). Instead, a non-classical psychology takes into account the full variety of relations that may hold between object and method in terms of the observer's capacities to distinguish the two. Predominant in this arrangement is the central asymptotic 'distinguishability' value, at which 'priority' is undefined. As a limit case of the outside tradition, the method has become here its own object of investigation. Likewise, as a limit case of the inside tradition, the object has become here fully its own method of investigation. Thus, at this point a fully self-referential relation occurs between object and method, that we may consider as a merge between the two. It is by accepting this point that the implicit priority assignments, which distinguish the two classical traditions, can be overcome.

By including this point and its vertical asymptote we admit a basic incapacity for an observer to maintain *always* a clear object-method distinction. The two 'classical' traditions, then, can be

[3]The reader may wonder why we do not speak more generally of 'classical science'. It is because modern physics, par excellence, has been already non-classical since the twenties of this century, whereas current mainstream psychology still mirrors itself to 19th century classical physics.

considered as unwarranted generalizations of a perfect object-method distinguishability. They both leave out the undefined point, at which the self-referential merge of object and method takes place.

Rather than denoting some 'temporary imperfection' in his method, an observer's failures and inabilities to make object-method distinctions are now to be understood of *substantial* interest to his field of study. Thus, a whole universe of new empirical phenomena opens up, that is concerned with:

a) properties of objects of investigation that do not allow a clear object-method distinction;

and

b) the ways an observer may become entangled in attempting to keep his mind clear about object-method distinctions.

In other words: a universe of phenomena in which the observer himself also enters as a participant (cf. von Foerster, 1981).

It is precisely this universe that is also the work area of practical psychology, in particular psychotherapy. For if anywhere, it is here that

a) clear distinctions between the topics that are discussed and the ways in which these topics can be known, often disappear;

and

b) practitioners favor a culture of self-observation and the use of oneself as a supreme instrument in the study of their therapeutic interactions (e.g. Reik, 1948).

The Inaccessibility of Psychotherapy to Research

The field of psychotherapy is qualified by a great diversity of styles and schools, with a highly competitive relationship between them. This situation posed the question which therapy was better, and why. Nothing seemed more obvious than a straightforward comparison of the therapeutic effects of the various styles and schools.

Unfortunately, however, despite enormous research efforts, these comparative studies did not reveal much that was of interest to the practicing therapist. If anything became clear out of these investigations, it was that the original evaluation research questions were underestimating the complexity of the subject matter. Futhermore these studies contributed to the rise of a wide cultural gap between researchers and therapists. We might characterize this gap in terms of the following features:

I. *Clarity versus vagueness in clients' problems*:
Researchers are mainly interested in clearly observable phenomena, by which therapeutic effects can be measured as differences between pre- and post-treatment scores. Therapists, on the other hand, show a much higher tolerance of vagueness. Indeed, many psychotherapeutic processes are processes of clarification of initially incomprehensible problems. Clear understanding of the relevant phenomena is often an outcome of the therapy, rather than a precondition.

II. *Independence versus dependence of therapists' measurement activities*:
Many social scientists cherish the ideal of a science of objective facts that exist independently from their being measured, as in nineteenth century physics. Psychotherapists, on the other hand, act in such a way that their assessment activities not only influence the patients, but also constitute the phenomena that are considered of importance.

III. *Explicit versus implicit knowledge of therapeutic procedures*:
Researchers are looking for techniques that can be formulated explicitly, preferably in terms of step by step procedures. Therapists, on the other hand, make extensive use of their intuitions, that, however, cannot be made fully explicit. Very often researchers are struck and stuck by the vagueness of therapists' accounts of their performances.

IV. *Control versus autonomy of therapeutic processes*:
Researchers are interested in therapeutic processes as developments that the therapist is controlling. Research aims at describing effective procedures. Therapists are much less concerned with control, and often consider therapeutic interaction processes as

spontaneous developments. Therapy aims at restoring individual autonomy.

Kaplan's (1964) "law of the instrument" states that the available measuring procedures may impose themselves upon the topics of study: "Give a small boy a hammer, and he will find that everything he encounters needs pounding" (p. 28). Accordingly, in a research culture dominated by classical ideas, psychotherapy is primarily investigated as a particular set of events, observable by the appropriate measuring devices. This position is taken particularly by those who study, and prefer to study, behavior therapy, since this style of therapy is considered the most 'scientific' and yielding the clearest therapeutic outcomes. But this should be hardly surprising. For as a therapeutic procedure it is dressed to fit in with the mainstream research culture. The greater ease of measuring its outcomes is mistaken for a higher degree of therapeutic quality; a tautology is welcomed as a contingency.

A widely practiced group of so called 'uncovering' styles, such as psychoanalysis, client-centered therapy, gestalt therapy, turn out to be too 'vague' or otherwise difficult to investigate by means of mainstream research. The research problems encountered are usually formulated as a lack of valid measures in respect of both outcomes and interaction processes.

A non-classical interpretation would hold that these uncovering therapies, as well as other, more 'researcher-friendly' styles of therapy, require a different way of investigating: one that doesn't require a strict distinction between observing and being observed, measuring and being measured. This would amount to a way of investigating in which the researcher recognizes the occurrence of object-method merges, both in his own research efforts, and in the perceptual and cognitive activities by therapists and clients. Then vagueness, measurement dependence, implicit knowledge, as well as the autonomy of therapeutic interaction processes become phenomena of substantial, and not of merely methodical, interest. This is the promise of self-organization.

One quasi-solution by researchers to overcome the differences between research and practice has been to declare many therapeutic styles 'unscientific', and to concentrate research efforts mainly on those therapies, which allowed the observation of maximally clear and unambiguous phenomena. This is particularly the case for behavior therapy, which has been studied extensively for that reason. It is a quasi-solution, because no proper research method was found to study the other therapeutic styles.

For that purpose, a fundamental revision of the research assumptions is needed, which amounts to a reformulation of the relevant features of psychotherapeutic processes. To contribute to such a revision is the aim of the present book. This will be done by using a conceptual frame that has been developed in cybernetics.

Second Order Cybernetics and Self-Organizing Systems

The works of Maturana, von Foerster, Pask and other leading theorists on self-organizing or autonomous systems have brought a change of attention in cybernetics. In regular 'classical' cybernetics the objects of study are systems that are defined as external to the investigator. They can be defined, observed, and computed. They are devices of control. Best known example is the thermostat. The performances of the investigator himself do not affect the system, only perhaps marginally and incidentally. The investigator does not partake in the definition of the system. In 'second order', or 'non-classical' cybernetics, however, this is principally different: the systems that are studied differ from classical cybernetics, in that the observer enters, in some way or other, into the very definition of the system. Or, in other words, the system studied obtains autonomous qualities.

Now if the observer's acts of observing are no longer external to the system studied, then the object of observation consists to a high degree of the very acts of observing this object. The study of self-organizing systems therefore primarily amounts to a formal theory of systems that are able to behave as observing, knowing, autonomous individuals. It is concerned with the specification of how these systems should be organized, and how observation, cognition and autonomy should be understood formally.

In this discipline we encounter the same kind of self-referential merge which we discussed above in terms of object and method. In autonomous, self-organizing systems a merge is established between that which is organizing and that which is being organized. A theory of self-organization is simultaneously concerned with both the organizing process (cf. method) and the organized outcome (cf. object). As a theory of perceiving and knowing it entails an epistemology; as a theory of autonomous the systems capable of perception and cognition it also entails an ontology.

Self-Organization in Psychotherapy

Therapeutic conversations often have themselves as the matter of discussion; therapists use themselves as supreme measuring instruments; clients explore their own ways of exploration. It is this self-referentiality which can be understood in terms of the emergence or the maintenance of self-organizing properties of therapeutic interactions (cf. Goudsmit, 1989).

Though many characteristics of self-organizing systems are widely hinted at in the therapeutic literature, a comprehensive theoretical framework is missing and current research traditions are not well adjusted to study them. It is the task of unraveling and understanding therapeutic processes according to the theories on self-organizing systems, which makes the latter a most exciting new development in psychology.

In the field of psychotherapy these developments in second order cybernetics have already been welcomed, so far, primarily by family therapists. They recognized a vital theoretical support for their common practice of abstaining from questions on 'truth' and 'reality' in family conflicts. But also for other styles of psychotherapy these developments are of importance.

In view of this, a symposium on self-organization in psychotherapy was organized at the University of Groningen, on may 9-11, 1988. All but one of the papers in this book have been presented there.

The present book offers some perspectives on psychotherapy, that have self-organization as a common theoretical foundation. The contributors to this book pay attention to different schools of psychotherapy: not only family and systemic therapy, but also gestalt therapy, personal construct therapy, art therapy and psychoanalysis.

In chapter 2 Vincent Kenny introduces some of the basic notions of a major modern theory of autonomous systems: Humberto Maturana's theory of autopoiesis. This chapter is a revised version of an invited paper presented on October 2, 1985 at the Istituto di Psicologia, Università Cattolica del Sacro Cuore, Rome.

In chapter 3 Gerhard Portele presents a comparison between gestalt psychology and Maturana's theory of autopoiesis. He shows how gestalt therapy can be considered as a practical elaboration of the latter, and how the notion of autopoiesis allows a more elegant description of the crucial features of gestalt therapy.

In chapter 4 Henri Schneider discusses psychotherapeutic interactions from the points of view of complex dynamics and order through fluctuations. Self-organization takes place through the emergence of self-referential experiences in patients during treatment.

In chapter 5 Vincent Kenny makes a comparison between Maturana's theory and that of George A. Kelly on personal constructs. Attention is given to various kinds of constructivism. Personal constructs are described as vital elements in self-organizing systems.

In chapter 6 Kim James presents some central notions from the work of J.J. Gibson on direct perception, and relates these to therapeutic processes, in particular to a psychotherapeutic technique which makes extensive use of painting and other graphic activities.

In chapter 7 Max van Trommel uses concepts from Bateson and Maturana to describe the systemic therapist's possibilities of using him/herself as an instrument of interventions and assessments, switching between a neutral and a non-neutral stance.

Finally, in chapter 8 I will close by making a few remarks on blindness and incomprehension in psychotherapy and research.

References

Armenakis, A.A. (1988). A review of research on the change typology. *Res. in Organ. Change and Development*, **2**, 163-194.

Bohr, N. (1934). The quantum of action and the description of nature. In: N. Bohr, *Atomic theory and the description of nature*. Cambridge: Cambridge U.P.

Foerster, H. von (1981). On cybernetics of cybernetics and social theory. In: G. Roth, H. Schwegler (eds.), *Self-organizing systems. An interdisciplinary approach*. Frankfurt M./New York: Campus.

Goudsmit, A.L. (1989). Organizational closure, the process of psychotherapy and the psychologist's fallacy. In: G.J. Dalenoort (ed.), *The paradigm of self-organization*. New York: Gordon & Breach.

Holton, G. (1973). *Thematic origins of scientific thought*. Cambridge, Mass.: Harvard U.P.

Kaplan, A. (1964). *The conduct of inquiry. Methodology for behavioral science*. New York: Harper & Row.

Kelley, H.H., E. Berscheid, A. Christensen, J.H. Harvey, T.L. Huston, G. Levinger, E. McClintock, L. A. Peplau, D.R. Peterson (eds.) (1983). *Interpersonal relations: a theory of interdependence*. New York: Wiley.

Merleau-Ponty, M. (1945). *Phénoménologie de la perception*. Paris: Gallimard.

Polanyi, M. (1966). *The tacit dimension*. New York: Doubleday.

Reik, Th. (1948). *Listening with the third ear. The inner experience of a psychoanalyst*. New York: Grove Press.

Shotter, J. (1985). Accounting for place and space. *Environment & Planning D: Society & Space*, **3**, 447-460.

Shotter, J. (1987). The social construction of an "us": problems of accountability and narratology. In: R. Burnett, P. McGhee, D. Clarke (eds.), *Accounting for personal relationships: social representations of interpersonal links*. London: Methuen.

2

Life, the Multiverse and Everything; an Introduction to the Ideas of Humberto Maturana

Vincent Kenny

Abstract. This chapter introduces the central concerns of Humberto Maturana's theory of autopoiesis as they relate to the domain of psychotherapy. Several common terms which are redefined within his theory in an unusual manner are unpacked as to their idiosyncratic significance including the expressions, 'linguistic behavior', 'languaging', 'structure determinism', 'organization', 'structure' and others. The source material used for this exposition include not only the cited texts but also several workshops from which verbatim transcripts are often used in the form of brief quotations. I have attempted to stay as close to the original material as possible in order to convey both the meaning and the texture of Maturana's work. This is not an easy theory to grasp, ranging as it does across several specialist fields from the neurophysiology of perception through social communication to epistemology. Nor are the implicative transitions from a theory of biology to the praxis of psychotherapy without complexity and controversy. Nonetheless, Maturana offers a novel theory of conversations which could form the basis of a much needed new paradigm for personal change.

Life: Love and Languaging

MATURANA USED TO use the phrase 'biological stickiness' to describe how any two systems, upon encountering one another, stayed or 'stuck' together. They fit together and remain together and continuously interact recurrently with each other. More recently he has used the more dangerous word 'love' to describe this happening of living. Love is a phenomenon which takes place a priori, without precedent, and without prior justification. Maturana claims that if you tell someone that: "I love you because you are so beautiful/ intelligent etc." then either you do not really love that person or you are pretending to have reasons for something for which there are no reasons. 'One simply falls in love and every love is love at first sight even if it arises after living together for 20 years.' In other words love is an expression of a particular structural configuration in the two participants such that they stick together with no reason. Love is a primary constitutive condition and is fundamental if social phenomena are to arise.

Being in love means making a space for one another so that each becomes part of the domain of existence of the other, and within their continuous recurrence of interactions they form a system in which they have a co-ontogeny. It is the recurrence of interaction within the medium that creates the conditions for co-ontogeny. If they fit, one with respect to the other, then they form a path of (structural) drift together. Within this co-ontogenic drift new phenomena will arise immediately.

Without love there would be no social phenomena. This is an important point since for Maturana many crucial human phenomena are social e.g. language, self-awareness, mind, self etc. By ontogeny is meant the living system's history of structural drift in which its course of structural changes is contingent upon the interactions it undergoes in its medium. Each interaction triggers a particular change and the next interaction triggers another particular change and so forth.

The living system and its medium are operationally independent and so whatever changes of structure take place are determined by the structure of the system itself at every moment. The path of change is contingent upon the history of interactions in the medium. When we look retrospectively we can see that the system and the medium are in correspondence, i.e. they are in congruence with one another.

"*Every system is where it is, in a present, in congruence with its medium, and cannot be anywhere else.*"[4] This is a typical statement by Maturana whereby he means to underline the coherence and congruence of each system in its domain of existence. A human system may not like where he is in the medium, and may feel extremely badly about what "life" has doled out to him, but he is where he is through a coherent series of structural interactions and changes in his ontogenic drift. It is interesting that we apply the word "drifter" in a pejorative manner to those folks who most obviously exemplify the human condition of structural drift, as if we, by our 'rootedness' were escaping this essential constraint and thereby exerting 'control' or 'steering' over our lives in a determining way.

Both the living system and the medium change in congruence with one another. They change their structure/shape so that they fit together in a drift. The concept of drift does not imply a chaotic situation because it is being determined on a moment-to-moment basis by the interactions. The path of drift is contingent upon the interactions. So unilateral steering is an illusion. This path of drift is a path without any choices. It is a path of conservation of (a) the organization of the living system and (b) of congruence with the medium. This is the paradigm for survival.

When we have two living systems (A and B) interacting with one another each one forms part of the medium of the other. Within their co-ontogenic structural drift A's structural drift is contingent upon its interactions with B in the medium and vice versa. From an observer's point of view you could describe this co-ontogeny as the co-ordination of actions between A and B, since there are consequences for A/B of each others actions in the medium. Further, we can say that without this co-ontogeny, certain

---

[4]As mentioned in the abstract above, these quotes without references are verbatim material from workshops conducted by Maturana in 1985.

behaviors between A/B would not have arisen. Within the co-ontogeny the behaviors of A/B become consensual - i.e. they have created a consensus about the coordination of their behaviors).

Consensual behavior is behavior between two systems as a result of living together These would not have appeared had they not lived together. The behavior is contingent upon their ontogenies. These behaviors can be described as interactions in the medium. This consensual coordination of behavior is what Maturana calls *linguistic behavior*. Examples of linguistic behavior can be easily observed occurring between humans and their pets. One instance being the consensual coordination of a cat scratching a door to be let out of the house by its owner. Another may be seen if you inadvertently move to stand on your dog's tail, it moves its tail out of the way of your foot. These behaviors arise because of their co-ontogeny, of living together.

An observer could describe these interactions in semantic terms, i.e. one could ascribe meanings to the elements in the coordination of behavior e.g., "the cat is telling his owner that he wants to go out". However Maturana is keen to point out that there is no intrinsic meaning in the linguistic behavior. What is happening is that the two systems (person + cat) trigger various structural changes in one another. Maturana gives the following example of structural changes triggered by interactions to underline the absence of meaning: in the process of lens-making two pieces of glass are ground together. By using certain rotating movements you will produce one concave and one convex lens. We could either say that these two fit together or that the concave is meant to contain the convex.

However, this ascription of meaning (of purpose/intent) is not a feature of the geometrical correspondence. What we have are changes of structure contingent upon their interactions. We have two congruent structurally dynamic entities such that the changes of structure of one trigger congruent changes of structure in the other which in turn trigger changes in the first which are congruent with it. By sticking to structural descriptions Maturana aims to empty out all other types of symbolic explanations. The starkness of Maturana's position is ameliorated by Varela (1981) who, while agreeing with Maturana that notions of purpose, information or code cannot play any logical role in the description of autopoietic systems, points out that our human cognitive capabilities will remain unsatisfied unless such explanations are also complemented with carefully constructed symbolic explanations.

The coordination of action in relation to interactions in the medium is called Linguistic Behavior or Linguistic Interaction. This always takes place when two living systems live together and have structural plasticity in the domain of their recurrent interactions. Structural plasticity is necessary, in that the systems must be able to change their structures when triggered by one another.

> "*The plastic splendor of the nervous system does not lie in its production of 'engrams' or representations of things in the world; rather, it lies in its continuous transformation in line with transformations of the environment as a result of how each interaction affects it.*" (Maturana & Varela, 1987, p. 170).

Given sufficient structural plasticity and the continuation of recurrent interactions then we may observe the coordination of behavior - not only in relation to interactions in the medium but also in relation to these coordinations of actions. That is they coordinate their behavior in relation to the coordination of behavior. We observe consensual behavior about consensual behavior. We see linguistic behavior about linguistic behavior. This is what Maturana calls "Language".

When we get a recursion in the coordination of consensual behavior, so that there is consensual coordination of behavior of consensual coordination of behavior then we have this new phenomenon which is language.

> "*So, we can also say that language is a domain of recursive linguistic co-ordinations of actions, or a domain of second-order linguistic co-ordinations of actions. We human beings also co-ordinate our actions with each other in first-order linguistic domains, and we do so frequently with non-human animals.*" (Maturana, 1988, p. 48).

For Maturana several important phenomena arise with language including:
a) The Observer
b) Humanity
c) Meaning
d) Self-awareness/consciousness
and
e) Objects

What makes us human is languaging. "*Humanity arises in the social dynamics in which languaging takes place*". This is difficult to prove but Maturana cites examples of feral children brought up by wolves so that what we find are wolves with the genetics of homo sapiens. They never learn to speak (although they may know a few words).

It is important to note that no particular behavior or movement or gesture or sound constitutes languaging. Rather, it is an ongoing process because it is defined in the history of the coordination of actions. Just a word or gesture does not constitute languaging. Furthermore, languaging is not an abstract phenomenon, we are not dealing with abstract entities.

Languaging becomes part of our medium and so anything we say is not trivial since it becomes part of the domain in which our co-ontogenic structural drift takes place. That is, our co-drifts are contingent upon our languaging. Languaging interactions are as powerful as a physical interaction e.g. pushing someone hard. If I say "How beautiful you look" - this has certain consequences in terms of a "particular configuration of structural perturbations." This statement is like a caress. Equally, if I say "you look terrible" this is another particular configuration of structural perturbation. Such an interaction Maturana calls "like hammerings in the head", i.e. it is painful.

> "*Thus we say that the words were smooth, caressing, hard, sharp, and so on: all words that refer to body touching. Indeed we can kill or elate with words as body experiences. We kill or elate with words because, as co-ordinations of actions, they take place through body interactions that trigger in us body changes in the domain of physiology.*" (Maturana, 1988, p. 48).

Structural changes triggered here include changes in blood pressure, blood flow, hormone flow production, brain synapses undergoing different changes etc, all depending on what is said. These changes take place unavoidably as a process of structural change contingent to the interactions and hence as a drift because the course of structural change is being specified on a moment-to-moment basis in the interaction.

However, **drift** will only go in the direction that the circumstances will allow. Drift will not go in any imaginable direction. The example here is to consider the path of a boat which has no rudder, oars, engine, or mast etc. being generated as a drift. Even if we could specify and compute the structure of all the systems involved and were thereby able to predict the direction of the drift (which we cannot do) it would still be a drift, because the system flows in its own dynamic of structural changes. This is not to say that we cannot alter the direction of the drift for example by what we do in languaging since this (languaging) defines conditions in which the drift takes place. If we language one way ("you're beautiful") the drift goes this way rather than that way ("you look terrible"). The human dilemma is that we want to pretend to control our lives (and others' lives) as if we could specify the outcome of the drifting pattern.

The notion of control arises in the context of productivity. Maturana talks about 3 main modes in which we can act and these 3 are distinguished largely in terms of differences of intent.

Firstly, the Science mode - the intent here is explanations.

Secondly, the Technology mode - the intent here is production.

Thirdly, the Art mode - where the intent is aesthetic.

Within the Science Mode the approach is to introduce variety in order to be able to generate more comprehensive explanations of phenomena. That is, novelty is introduced as a means to an end.

The Art Mode is to amplify free creativity to generate a self-saying aesthetic phenomenon which needs no further explanation. The artistic piece is self-producing and self-sufficient in its final form. Here novelty is produced as an end in itself.

In the Technological mode we intend to achieve a particular result and so we specify certain constraints on the variability of the components of the system, with the result that the drift can follow only one particular course. Here novelty is excluded by systematic controls. This applies equally to technological supervisors in a car factory (ensuring that each car is produced with minimal variation), as well as to fascist dictatorships whose technological supervisors serve to control and eliminate any dissenting voices.

Although we cannot control our co-drift since its path is formed by moment-to-moment interactions, and although the concepts of choice and free will become redundant in this regard, we must still be extremely careful about our actions since whatever we do forms part of the medium in which we drift, and therefore we drift in a different way to how we would drift if we did nothing. So what we do is not irrelevant to our drift, even if we cannot actually control the dynamics of the drift mechanism. Thus, whatever we do in languaging is not trivial because languaging is a manner of moving in a co-drift which makes it possible for us to complexify our human lives together.

The amount of complexity we can generate in human behavior in terms of the recursion of coordination of actions about the coordination of actions is open or infinite. But nothing that we do in it is trivial. *"All that takes place in human life is languaging, and all that takes place in languaging is conversations". "These are continuous mutual grooming interactions. We immerse ourselves in structural drift contingent to the conversations in which we participate and which we generate through our structural dynamics".*

Note that language does not take place in the brain but rather in the social dynamics. Languaging is a way of being together in a collective, it is a way of co-ontogenically drifting. [Without the brain there is no language, but language does not exist in the brain]

Self-consciousness arises in languaging as a manner of consensual coordination of distinctions about the consensual coordination of distinctions in which the participants (i.e. those who are distinguished) are distinguished. In languaging we can reflexively describe ourselves, and describe ourselves describing ourselves and so forth. We do this through linguistic distinction of linguistic distinctions.

> "*Self-consciousness arises in language in the linguistic recursion that brings forth the distinction of the self as an entity in the explanation of the operation of the observer in the distinction of the self from other entities in a consensual domain of distinctions.*" (Maturana, 1987, p. 375).

So we see that self-consciousness depends upon languaging as a phenomenon of linguistic recursion. Self-consciousness, self-awareness, and mind are social phenomena because they take place in languaging, in the social domain.

Another importance of language within Maturana's system is that prior to language there are no objects. That is, objects arise with language. Objects are entities specified in the coordinations of coordinations of consensual actions.

> "*...the participants of a consensual domain of interactions operate in their consensual behavior making consensual distinctions of their consensual distinctions, in a process that recursively makes a consensual action a consensual token for a consensual distinction that it obscures.*" (Maturana, 1987, pp. 359-360).

What this means is that the object we bring forth obscures the operation of distinction it stands for. When I use my pen to ink marks onto this piece of white paper, the action of writing or 'inking' is an operation of distinction whereby I bring forth the inked words on the page. So 'inking' as an action is an operation of distinction, and my inked words may become consensual distinctions as soon as I coordinate my actions with other persons. When consensual distinctions obscure the actions they stand for, then they become objects.

So objects arise in languaging and at the same time obscure the operations of distinction for which they stand. Hence we are left with these entities which seem to exist independently of everything. This illusion of independent existence is achieved because the objects obscure the operations of distinction that constitute them. In this way objects are reified. "*In the recursion of consensual distinctions of consensual distinctions we continually transform notions/ concepts into objects"*.

Prior to human beings there were no objects, since objects arose with language. If we see a cat chasing and catching a mouse, then for Maturana the 'cat' is (not) eating the 'mouse'. Rather *"it is flowing in the structural dynamics of its structural coupling/congruence in its domain of existence"*. The 'cat' does not exist as a 'cat' for the 'cat'. It cannot exist until somehow language arises for the cat.

"We humans also 'do' many things without doing them. We 'walk' without walking. We perform many actions which we can talk about afterwards, but which do not pertain to the domain of languaging while we are performing them. So we are not doing them."

Many of Maturana's ideas, including the distinction between the domain of experience and the domain of explanations, and the impossibility of instructive interactions because of the structure-determined nature of living systems, can be read as echoes of Lao Tsu's work "Tao Te Ching", as the following passages from the Tao illustrate.

"A truly good man is not aware of his goodness,
And is therefore good.
A foolish man tries to be good,
And is therefore not good.

A truly good man does nothing,
Yet leaves nothing undone.
A foolish man is always doing,
Yet much remains to be done."

"In the pursuit of learning, every day something is acquired,
In the pursuit of Tao, every day something is dropped.
Less and less is done
Until non-action is achieved.
When nothing is done, nothing is left undone.
The world is ruled by letting things take their course.
It cannot be ruled by interfering."

In his workshops of 1985 Maturana would say that we are not in languaging all the time, referring to the previous comments of doing things without doing them. One intention is to distinguish between the two non-intersecting domains which he calls the Domain of Experience and the Domain of Explanation. In postulating that we can never have less than these two dimensions Maturana claims that his approach is not reductionistic.

While, as observers, we are all in languaging all the time, language is not the only means we have of operating in consensual co-ordinations of actions. As we have seen, prior to the development of language are the linguistic co-ordinations of actions. So the decision as to whether or not we are in languaging when we are alone depends on whether or not the actions we are undertaking belong to some implicit domain of consensual co-ordinations of actions within our observer community. With this in mind we can understand that certain individuals are called 'mad' or 'eccentric' because they are seen to be enacting languaging but outside of any implicit or explicit domain of consensuality.

Conversations as Structural Perturbations

As humans we dwell in language, and are realized in the social domain through languaging, through our constitution of conversations in which we bring forth objects as if they were fixed entities. It is as if these objects exist independently of any observer (i.e. we assume that we 'discover' reality).

> "*In daily life we call conversation a flow of coordinations of actions and emotions that we observers distinguish as taking place between human beings that interact recurrently in language....the different systems of co-existence, or kinds of human communities that we integrate, differ in the networks of conversations (consensual coordinations of actions and emotions) that constitute them, and therefore, in the domains of reality in which they take place. Emotions are not conversations, but we flow in our emotioning through the flow of our conversations.*" (Maturana, 1988, p. 53)

Not all conversations elicit emotions, as we know. Maturana outlines a (non-exhaustive) list of six classes of conversations which we can distinguish among human interactions. These are defined in terms of differences in the pattern of coordinations of actions and emotions which are variously invoked and are as follows:

1. *Conversations of coordinations of present and future actions.* Such conversations are used for the actual coordinations of actions which take place in relation to a particular domain. The conversational participants are only contributing to the coordinations of actions and there is no particular emotional content.

2. *Conversations of complaint and apology for unkept agreements.* These coordinations of actions, within the frame of emotions of righteousness and guilt are concerned with demands and promises.

3. *Conversations of desires and expectations.* These are coordinations of actions undertaken by participants whose attention is oriented to future descriptions and not to the current actions through which they are being constituted as humans in the present.

4. *Conversations of command and obedience.* Such coordinations of actions take place within an emotional frame of negation. That is, by complying with commands to do as he otherwise would not do, the one obeying the commands both negates himself and the person commanding (by attributing to him a characteristic of 'superiority'). The one commanding also engages in this dual negation.

5. *Conversations of characterizations, attributions and valuing.* Here the coordinations of actions are embedded in an emotional flow of acceptance and rejection, together with the experience of pleasure and frustration depending on whether or not the listeners feel they have been correctly recognized or not by the speakers.

6. *Conversations of complaint for unfulfilled expectations.* In this case the listener feels frustrated by being accused of not fulfilling a

promise that he did not make, while the speaker feels frustrated that the listener has dishonestly not kept a promise made.

> " ... *as we human beings participate in many different conversations simultaneously or in succession, our actual community coexistence courses as the changing front of a network of conversations in which different crisscrossing coordinations of present and future actions braid with different consensual emotional flows.*" (Maturana, 1988, p. 53)

By emphasizing the interweaving of languaging and emotioning, Maturana unpacks further his notion that conversations are structural perturbations which have far-reaching effects on our bodyhoods. Our 'self' or 'identity' is defined by the totality of all the systems of social interactions in which we participate. In this sense our bodyhood is the time/space location of structural intersections of the many different systems which we constitute or participate in bringing forth through our actions.

Anyone familiar with John Searle's (1969) work on speech acts will note a strong similarity between these conversational types outlined above and Searle's five categories of illocutionary point, namely, assertives, directives, commissives, expressives and declarations. These outline varying patterns of commitment coordinated by speakers and listeners.

To close this section on conversations it is important to recall that languaging does not denote independent objects, but is rather a system of orienting behavior whose function is to generate a consensual domain of actions. It is to orient the listener within his own cognitive domain.

The Multiverse: Expanding the Universe through the Ontology of the Observer

Problems with Perception: How is it that We Make Mistakes?

It is constitutive for Maturana that at the moment of experiencing we cannot tell a perception from an hallucination. From his analysis, the science of neurophysiology has failed to generate a mechanism which could explain our hearing/seeing objects external to us, independent of us. Maturana asks: *"How come we make mistakes in perception if it is the case that we directly see an objective reality?"*. He points out that at the moment of perceiving we never know that we are making a mistake - this awareness of a 'mistake' is always post-hoc. It is only afterwards that we can say it was an illusion, hallucination or mistake. These two are indistinguishable in the experiential domain. Hence the differentiation of a perception from an illusion is a social distinction formed in consensus with others, (usually in conjunction with some authority who has an instrument). We believe the external source of authority. "Illusion" is seen therefore as an explanatory principle to 'explain away' a distinction which is experientially impossible. Social confirmation does not constitute proof of an independently existing reality.

In other words, if we take seriously the fact that in the experiential domain this distinction is impossible, then it follows that we cannot rely - for the validation of our arguments - on any assumption that entails having a privileged or direct access to 'outside' objects. The external object cannot be the source of validation for what we say. Hence, Maturana sees the assumption of an objective reality as a "miss-take", i.e. erroneously taking as independent of us entities which we ourselves bring forth. The willingness to make this miss-take he finds to be based on a search for certainty. However he warns that *"certainty blinds, the more certainty the less you see"*.

Instead of certainty we need social coherence. This is for example what science is. Every ideology, game, club etc. is a domain of social coherences defined by the consensus criteria for acceptability of statements.

> *"Coherence and harmony in relations and interactions between the members of a human social system are due to the coherence and harmony of their growth in it, in an ongoing social learning which their own social (linguistic) operation defines and which is possible thanks to the genetic and ontogenetic processes that permit structural plasticity of the members."* (Maturana & Varela, 1987, p. 199)

In abandoning the certainty of having a privileged access to objective reality Maturana puts objectivity into parentheses, thus: (objectivity). In this way we have two very different explanatory pathways which he refers to as:

a) The path of objectivity without parentheses (= the way of the Transcendental Ontologies),

and

b) The path of objectivity in parentheses (= the way of the Constitutive Ontologies).

In the first way the observer assumes that existence takes place independently of what he does, that objects have independent and separate existence, and that these can be known directly through processes of perceiving and reasoning. The criteria for acceptability of the truth of statements refers to some independently existing source of validation (e.g. God, rationality etc.).

This way of explaining necessitates the observer to further assume a single reality, a Universe (the Transcendental referent) which is the source of validation for all explanations, and hence for the way we explain our praxis of living. Disagreements among competing observer explanations necessarily involve claims of privileged access to what is 'really real' and consequent mutual negation.

In following the second path the observer assumes, quite differently, that the starting point must be the constitutive biological phenomena of being unable to distinguish perception from illusion in daily living. In the absence of being able to make statements about independently existing objects to which one has privileged access, this pathway focusses on the ontology of the observer, on what the observer does to bring forth objects in a domain of existence through consensual operations of distinction. The criteria for acceptability of statements shifts therefore to observer community agreements and away from objectivity. Both 'objects' and 'domains of existence' depend upon the observer. Thus the observer is the source of all realities and existences and can bring forth many different legitimate domains of reality through the operational coherences of his praxis of living.

While the universum is the ultimate reference cited for the validity of any statement in the transcendental path, the Multiversa is entailed by the parenthetic path, and implies that a multiplicity of realities can be brought forth depending only on the distinctions of the observer.

> "*Each versum of the multiversa is equally valid if not equally pleasant to be part of, and disagreements between observers, when they arise not from trivial logical mistakes within the same versum, but from the observers standing in different versa, will have to be solved ... through the generation of a common versum through coexistence in mutual acceptance. In the multiversa coexistence demands consensus, that is, common knowledge.*" (Maturana, 1987, p. 332).

The social consequences of both positions are completely different.

At this point it should be clear that for Maturana there is no objectively existing reality. Whatever reality we experience it is one of our own creation, i.e. we bring it forth through our operations of distinction. For Einstein, scientific theories were seen as the free creations of the human mind which we used to explain the world - but for Maturana, what needs explaining is precisely this "free creation of the human mind", i.e. the way in which the observer brings forth his world. Thus, central to Maturana's theory is the ontology of the observer. "*Languaging takes place in the happening of living of the observer. To explain languaging, I must explain the living of the observer*". Languaging is therefore Maturana's instrument for explanation and also his central problem.

Operations of Distinction

Q. What is an 'Observer'?

A. An observer is any being who can be in language speaking with another (or to himself) and making distinctions.

Q. What does an observer do?
A. He makes distinctions.

Q. What is a distinction?
A. Any operation that we may enact which results in the separation of an entity from a background, i.e. which simultaneously distinguishes a unity in its domain of existence. Thus we see that the existence of all phenomena is brought forth through making the appropriate operations of distinction. For example, I may bring forth a chair by making the operation of distinction of 'sitting-down'. To give another example, if we want to know how many people there are in a room we will make the operation of distinction of counting them.

We may distinguish two types of unity or system, namely a Simple Unity and a Composite Unity. When we bring forth a Simple Unity we bring forth an entity characterized as separable from its domain of existence in terms of its properties. It is totally characterized by its properties which distinguish it from its background, [i.e. we don't analyze it or decompose it]. Its properties are the dimensions that specify or characterize its distinction from all else. These properties arise because they are constitutive.

With the composite unity we do something more. Firstly, we distinguish a simple unity and then we decompose it and separate its components and relations. In the Universe we would claim that the composite unity composed itself, independently of us and what we do. We would assume that the components were either there or not there, and that its characteristics were intrinsic, inherent and eternal. However, in the multiverse it is we[5] who separate out the

[5]Much of the aesthetic and constructivist concerns here can be seen in the writings of Vico whose 'verum-factum' principle - that what is true is what we ourselves have made or constructed - went alongside his vision of the nature of the human mind as that
"*...which rejoices in the highest degree in that which forms a unity, comes together, falls into its proper place; ... that just as beauty is the due proportion of the members, first each to each and secondly as a whole, in any outstandingly lovely body, so knowledge should be considered as neither more nor less than the beauty of the human mind...*" (Vico, 1732, p. 239).

components and when we do this, we find that the components we bring forth have a peculiar relationship with the simple unity that they integrate, i.e. we say that the "whole is greater than the sum of its parts"[6]. Maturana finds this expression somewhat obscure because it does not reveal what this "greater" is which is brought forth.

What is meant is that the composing of this unity takes place in a very peculiar and particular manner and that this is to do with the relations which the components must hold between them so that they constitute the original simple unity that we have decomposed.

Complementary Relationship between Components and Unity:
Note that the components are components only to the extent that they compose the composite unity. That is to say, a component is a component only as a component. There are no free (spare) components hanging about the world. Nothing is a such a component. Something is a component only in composition. In composition the relation between components and the unity that they compose is always unique - they are complementary. It is to this that Maturana refers when he makes his distinction between the Organization and the Structure of a system.

Composite unities have Organization and Structure. "Organization" refers to the manner of composition that defines the unity. "Organization" refers only to the relations between components that must always be present so that the composite unity will be a unity of a particular type. "Organization" refers to those relations which when present identify the unity as a particular type. Hence, the Organization of a system is necessarily invariant because if you change it you create something else. If the relations that constitute the unity change, the identity of the unity changes.

Forgers understand this principle very well because in trying to present a painting as a 'Renoir' what they do is to carefully maintain as invariant as possible (as resistant to scrutiny as possible) those critical relations (brushstrokes, texturing etc.) among

---

[6]See also the contribution by professor Portele in this book in regard of other similarities between gestalt theory and autopoiesis [ed.].

specified components (colors, oils, aged canvas etc.) which will identify it as that class of production called 'Renoir'. Experts attempt to distinguish between "fakes" and "the genuine article" by decomposing the artistic unity into its components and relations. The artist's "style" is that peculiar way in which he composed the constituent parts. The way he organized his painting. This Organization must remain invariant for the unity to conserve its class identity.

"Structure" refers to the actual components and the actual relations that realism a particular composite unity. While the organization is necessarily invariant (to conserve identity) structure is not. Structure is in continual change. Structure entails many more dimensions, more relations than organization. Organization can therefore be seen to be a subset of Structure. The Organization is always realized through Structure. We all structurally change continuously in our co-drifting. Living is a structural drift and lasts as long as Organization and correspondence with the medium is conserved.

Whenever we have a composite unity we have an organization that defines that unity as being of a particular class and we also have structure which refers to the actual manner in which that particular unity is material-ized.

There are two types of structural changes possible:

I	II
Changes where the organizational invariance is conserved	Changes without conservation of organization.

A living system will last as long as its organization is conserved and as long as it can be realized in its domain of existence. The structure of a system specifies the following four domains:

1) Domain of changes of state; all possible structural changes which the system can encompass while at the same time conserving its organization.

2) Domain of disintegrative changes; all structural changes a system can undergo but where the organization is destroyed.
3) Domain of perturbations; all interactions which trigger changes of state.
4) Domain of disintegrative interactions; all perturbations which trigger destructive changes in the system (loss of organization).

Since systems are endlessly structurally changing these four domains are never fixed for all time but will change congruently with the changes of the system. Also, since there is this peculiar relation of composition between the components and the unity that they constitute, it follows that whatever properties that a composite unity has depends on how it is composed and hence depends on its Organization and Structure.

Further, since Organization is realized (or material-ized) only through Structure, it depends on the actual Structural configurations of that unity. So, Composite Unities are unities whose characteristics depend on their Structure i.e. depend on how they are made!

To summarize to this point we have the following:

1) The observer arises with languaging.
2) Languaging becomes part of our medium.
3) Our co-ontogenic structural drift is contingent on languaging.
4) Languaging interactions are powerful perturbations.
5) These trigger structural changes.
6) We cannot control or predict our structural drift.
7) Prior to languaging there are no objects.
8) Objects obscure the operations of distinction they stand for.
9) Objects appear therefore to exist independently of our operations of distinction, of our bringing them forth.
10) It is constitutive that we cannot distinguish an illusion from a perception.
11) The central issue for Maturana therefore is the ontology of the observer.
12) We must move away from the delusory 'certainty' of the Universe to the freedom of the Multiverse.

To further summarize we also have the following:

1) By making operations of distinction we can specify simple unities and composite unities.
2) A Composite unity may be decomposed into distinguishable components.
3) Such components only exist as components to the extent that they compose the composite unity.
4) There is a particular relation of composition between the components and the unity they constitute.
5) This concerns the relations that must obtain between the components in order to constitute the simple unity.
6) These relations are the Organization of the system.
7) This is distinguished from the Structure of the system which refers to the actual components and their actual relations which realism the organization.
8) Organization is necessarily invariant, while structure continually changes.
9) There are two types of Structural change: Firstly: "Changes of State" which conserves organizational invariance. Secondly: "Destructive Changes" which destroys the Organization.
10) The characteristics of a Composite Unity depend on its Structure.

Brief Example: The Family as a System

For any system there are four initial questions which we may ask:

1) What type of System is it? How is it defined?
2) What is its Organization and Structure ?
3) Am I interacting with it as a Simple or Composite unity? If I am in the domain in which the system is a simple unity, I interact with the system through its properties as a totality, as a simple unity. However, if I am dealing with a composite unity I can only interact through the properties of the components.
4) In what ways can I interact with the structure so that I may trigger some change which will either conserve the organization or will destroy it?

To begin answering these questions in relation to family systems we see that, according to Maturana, families exist as simple unities in a peculiar domain, i.e. a social-descriptive domain. This is a domain in which we do not move or structurally couple. Therefore we interact with the family as a Composite unity, only through its components (members).

To further elaborate our beginning questions we look to which relations among these components define it as a family of a particular type, i.e. having a particular organization. The family organization brought forth as problematic must be disintegrated so that the members can do something different. So we must look for the network of conversations which contain the relations of constitution of the family. The only way to disintegrate the organization of the system is through interactions which do not pertain to relations of constitution of the system, but rather encounter the components (individuals, meaning systems) in an orthogonal manner (i.e. in a way that does not pertain to the constitution of the system). The way in which the family interacts with the therapist reveals their network of conversations and the interactions they enter into in order to constitute a certain type of system they call 'family'. That is, they reveal their constitutive relations. The complaints from family members arise out of the conflict between (a) the 'passion for being together' and (b) the negative emotions they trigger in one another. The only solution is to destroy one of these conditions. If the family wants to stay together then we must change the structures so that the recurrent interactions cannot continue. This means destroying the organization of the family as found in its networks of conversations.

Since any system must conserve its organization if it is to remain identifiably the same system, it is obvious that psychotherapy is essentially an anti-social enterprise geared to the destruction of invariance and traditions.

And Every Thing is Structure

By now Maturana's emphasis on Structure is clear.

1) Organization is realized only through structure.
2) All psychotherapy must be aimed at structural changes, since it is not possible to change organization directly.
3) The characteristics of a Composite Unity depend on its structure (how it's made).
4) Structure continually changes.
5) Drift is constituted by the moment-to-moment changes in structural interactions in the medium.
6) Languaging triggers structural changes.
7) Whatever happens during interactions depends on the system's structure.

This brings us to Maturana's notion of Structure Determinism.

1) Interactions in the medium only trigger structural changes of composite unities.
2) The structure of the system fully determines its interactions by specifying the variety of interactions it can undertake.
3) The structure of the system specifies what it will accept as an interaction and what will be ignored.
4) A major implication of these ideas is that "information" does not exist, and that instructive interactions cannot take place. You cannot by acting externally on a system specify what happens in that system.
5) You can trigger certain changes and you may know what will happen when you do this triggering by knowing the structure of the system but you cannot specify what happens in the system because that is specified or determined in the structure of the system.

Science can only deal with structure-determined systems, with composite entities, i.e. with systems whose structures determine what will happen. In proposing a generative mechanism as an explanation for the phenomenon to be explained science always proposes a structure-determined entity.

Since we, as living systems, are structure-determined entities, then whatever happens to us is determined by our structures and is never determined by whatever we encounter in our medium. It may be triggered by perturbations but not determined by them.

Maturana gives the example of hitting someone on the head with a hammer. It is not the hammer which determines that you will die, it is the thickness of your skull. If your skull was made of rubber, the hammer would simply bounce off. The notion of instructive interactions belongs in the Universe of linear causality. Maturana quotes the story of King Midas the man with the golden touch who had asked for this power of instructive interaction. That is, he could determine completely the structure of other systems (from the outside). Maturana points out that the tragedy of Midas was that he could not be an analytic chemist. Midas could not do science because to do science you must claim that the characteristics of the system you analyze depend on the structure of the system and not on what you do to it.

"*It is constitutive for science that we can handle only structure-determined systems and that instructive interactions do not take place*". This structure-determinism does not entail predictability. We are determined but not pre-determined. Determinism means that the structural coherences between systems are satisfied.

"*Co-ontogenic structural drift takes place as a structure-determined phenomenon because it takes place in the domain of structure of the interacting composite unities*".

Maturana defines a living system as one which must conserve both its organization (identity) and its means of adaptatio to its environment (structural coupling).

Thus the Autonomy of the living system is paramount. The system can only do what it does at any particular moment of doing. There are no other choices in the system. A system is always in its proper place and cannot be mistaken.

'Structural Coupling' is a term (like adaptation) which is used to refer to the system's structural correspondence with its medium. A structure-determined system is coupled to its domain of existence (medium) as long as its organization is conserved and also as long as it conserves its congruence with the medium. Survival therefore consists in the simultaneous twin conservation of class identity and adaptation. If one of these conditions is lost then at that moment so is the other one.

From all of this we can see that to speak of a living system implies:

1) That this system is a structure-determined dynamic system.
2) That its organization is being realized, and
3) That it is being realized in a domain in which it undergoes reactions which trigger only changes of state (perturbations) (i.e. which retains organizational invariance) and does not undergo destructive interactions.

Maturana characterizes the living as 'autopoietic', which he defines in this way:

> "... *a composite unity whose organization can be described as a closed network of productions of components that through their interactions constitute the network of productions that produce them, and specify its extension by constituting its boundaries in their domain of existence, is an autopoietic system.*" (Maturana, 1987, p. 349).

Autopoiesis is therefore a very particular type of organization characterized by a recursive self-production where it is impossible to distinguish the product, producer or production. It is this recursive self-production which constitutes the so called 'organizational closure' of the living system.

Implications of Maturana's Theory for Psychotherapy

I will conclude this introduction with some brief and general implications.

1. How should we as therapists describe ourselves and what we do? Since causality is ruled out by virtue of the impossibility of instructive interactions then we can no longer think of ourselves as 'change agents' who operate on others to directly change them. This

is in line with Kelly's ideas on psychotherapy as providing an experimental context within which the person can productively ask questions through actions and thereby reconstitute or reconstruct himself. Furthermore, people do not 'begin' to change just because they have arrived in therapy, people are in the flux of change continuously. We must get into a co-ontogenic structural drift with the client but we cannot control this drift. The structural changes which arise in all the persons undergoing a co-ontogenic drift have particular implications for the therapist who is not excepted from these unpredictable transformations.

2. 'The system can only do what it does', means that the system can only learn what it is set up to learn. Teachers, for example, are familiar with the difficulty of trying to give "information" or "answers" to children who have no "questions" about the issue which happens to have importance for the teacher. Whenever we say "I find it difficult to hold his interest" we are in this domain of answers without questions.

3. For those therapists believing that there is a 'right' way for the complainant to become and a 'right' way for therapists to behave in order to get him there, Maturana's concept of the miss-taken nature of independently existing entities, such as a 'proper way to behave', may be a reason to move towards the Multiverse. Within the (objectivity) of the Multiverse and the concomitant need to validate statements through criteria of social consensuality, we can no longer usefully speak of the 'reality principle' or 'reality testing', but must speak in terms of 'participation in the construction of consensuality'.

4. Equally, since we exist as multi-selves in multi-verses then there is no 'right outcome' for psychotherapy, since there is no 'natural nature' for us to reach or achieve. In effect the emphasis shifts from getting the client 'back on his feet again' to triggering movement within the client system. A shift from 'problem-solving' to active participation in the creation of the observer-community coherences and to changes in co-ontogenic structural drift.

5. Individual responsibility becomes the center of attention within the framework of 'everything said is said by an observer' and that 'no-thing exists without an observer'. That is, we are fully responsible for what we bring forth in our lives. Events have no separate existence apart from our distinguishing them in words and symbols.

6. Related to this idea of the organizational closure of the observer is the fact that everything is necessarily transference. (Hence it's not something to be 'cured'). While Kelly would say that an observer's constructions say more about himself than about the events he is describing, Maturana goes further and says that the observer's utterances can only be a commentary about the observers own organizationally closed system. The closure of the system determines everything through system structure.

7. For family therapy there is now even more of a dilemma to define what 'family' means. When a family presents for therapy there are as many families sitting in your consulting room as there are observer/participants. Each person will describe the family he brings forth differently. The therapist's descriptions/diagnosis is just one more set of observer distinctions bringing forth a system in a domain of existence. It is important to note that it is not the one (same) individual observer. Rather, each observer brings forth a different reality by his operations of distinction. From the therapist's point of view he will distinguish what he regards as the structural dynamics which are constitutive of the family and too which each member contributes from his repertoire of multi-selves. It is through the redundant dimensions (i.e. those not constitutive of the family system) that the therapist must orthogonally interact. Furthermore, we cannot have a "family problem" since the 'family' can't speak (since it has no mouth). It is the individual speaker who complains and who constructs (or invents, or brings forth) the 'problem' through his languaging. Thus the processes of languaging bring forth an 'object' which is the family problematique and which becomes solidified as 'a problem-family'. This object obscures the operations of distinction which all the family members coordinate in, in order to continuously re-create the problem. Thus the family (and all

families) have a problematique, i.e., an a priori set of starting conditions, which are never brought into question and which form the basis of the conversations which in turn material-ize the family organization.

8. Aetiology (in terms of specifying causes for the development of problems) becomes irrelevant since simple linear cause-effect statements can only be a trivialization of the person's entire history of co-ontogenic structural drift. Outside languaging "*there are no starts or stops, beginnings or ends, causes or effects*". Different observers, through different operations of distinction, will bring forth different 'pathologies'. 'Pathology' is in the eye of the beholder, who is an observer with specific intents, and who operates within the consensual confines of his own observer community. Thus there is no "cause" we can usefully "discover" for anorexia. Such a search must remain an attempt to be reductionistic regarding the anorexic's history (reducing it to a set of abstracted professional constructs or categories of explanation). Anorexia is the situation reached by the sum total history to date of her co-ontogenic structural drift.

9. Maturana's theory also indicates that we must abandon causal concepts such as those of the "purpose" of symptoms: the "function" of disorders: the "resistance" of this complainant etc. All of these are attributions of an observer. As Kelly pointed out the notion of 'resistance' has more to do with the puzzlement of the therapist than with the obduracy of the client.

Concluding Comments

One of the greatest concerns in psychotherapy is how to generate enough space for the creative positioning of experiential explorations within the domain of therapeutic conversations, so that it becomes possible for the client system to spontaneously produce novel experience inconsistent with the frame of the client's problematique. From Maturana's theory we can abstract three very potent constructs which allow the therapist to do exactly this. While their abstraction from his theory is easy, their implementation is not and many therapists need to continually have observers of their conversationalist praxis with clients in order to successfully enact these three principles for the generation of space for novelty. These three are as follows:

Orthogonality: To be orthogonal means to interact with the client system in such a way as to not become enmeshed in the existing organization of the system as yet one more constituting component. When this occurs you become part of the problem and hence cannot be part of a solution. Acting orthogonally means selectively interacting with peripheral structure, i.e. components which are not actively involved in the constitution of the organization. The 'alien' nature of the therapist as a provoking stranger must therefore be conserved. Orthogonality is focussed primarily in the experiential domain where the individual refuses to intersect in relations of constitution of the problematique of another individual.

Parenthesizing: This clearly emerges from Maturana's theory concerning the ontology of the observer and underlying the fact that no objective reality exists independently of some observer. This view moves us to bracket or parenthesize all speaking and listening, all explanations, all descriptions, and to keep in the foreground the phenomenon of how objects come to obscure the operations of distinction of some observer who enacts these from a particular vantage point and with a particular intent. Widescale opportunities for the application of this principle can be found in the domain of referential objects, i.e. those objects (facts) which constitute what Waddington (1977) humorously called C.O.W.D.U.N.G., namely, the Conventional Wisdom of the Dominant Group. In other words, to be parenthetic is to deconstruct the unquestioned and apparently unquestionable reality of a given observer community.

Circularity: This principle we may derive from Maturana's emphasis on structure-determinism, and on the circularity and recursiveness of all organizationally closed systems. This moves us away from simplistic linear cause-effect sequences and towards the generation and appreciation of complexity and autonomy. Thus the elaboration of the original complaint-complainants network of conversations is conducted by many family therapists using a method of 'circular questioning'. This obviously occurs in the domain of conversational interactions. There are several complex aspects to the application of these three principles, but to select one of the most important as my final comment here I will say the following. Acting in orthogonal, parenthetic, and circular modes can be seen to be a necessary approach to deconstructing various forms of authority to which we otherwise make ourselves subject, and thereby help to generate obscurity and constriction within the three domains. Our possible personal experiential space, our actual social conversational interactions, and our unquestioned reality-fabric can, and must, all be radically elaborated, and transformed by a thoroughgoing involvement and embodiment of the notions of being experientially orthogonal, conversationally circular, and referentially parenthetic.

Orthogonality asserts individual autonomy and simultaneously brings into question the problematique of another person. Circularity asserts system autonomy (the larger whole) and simultaneously questions simplifications and the notion that some one person has the authority or vision to really know best what is going on. Parenthesizing asserts the creative autonomy of alternativism and simultaneously questions and undermines the invariance of pre-emptive past laws or rules for specifying reality. All three expand the flexibility of each domain and the possibilities of what may transpire within each. Finally, we may note a correspondence between the experiential domain and structure-determinism, the conversational domain and the Multiversa, and the referential domain and the ontology of the observer.

References

Lao Tsu (1972). *Tao te Ching*. London: Wildwood House.

Maturana, H.R. (1987). The Biological foundation of self-consciousness and the physical domain of existence. In: E.R. Caianiello (ed.). *Physics of cognitive processes*. Singapore, New Jersey, Hong Kong: World Scientific.

Maturana, H.R. (1988). Reality; The search for objectivity or the quest for a compelling argument. *Irish Journal of Psychology, Special Issue on "Radical Constructivism, Autopoiesis and Psychotherapy"*, V. Kenny (ed.), **9**, 1, 25-82.

Maturana, H.R., F.J. Varela, (1987). *The tree of knowledge*. Boston: New Science Library.

Searle, J. (1969). *Speech acts*. Cambridge: Cambridge University Press.

Varela, F.J. (1981). Describing the logic of the living. In: M. Zeleny (ed.), *Autopoiesis; A theory of living organization*. Oxford: North Holland.

Vico, G.B. (1976 (1732)). On the heroic mind. Published in: G. Tagliacozzo, M. Mooney, D.P. Verene (eds.). *Vico and contemporary thought*. Atlantic Highlands: Humanities Press.

Waddington, C.H. (1977). *Tools for thought*. Herts: Paladin.

3

Gestalt Psychology, Gestalt Therapy and the Theory of Autopoiesis

Gerhard Portele

Abstract. First I present what is called the "core" of gestalt psychology: the "law of natural order". It postulates a principle of self-organization. For Fritz Perls, the founder of gestalt therapy, the principle of "organismic self-regulation" is central to his theory of therapy. Furthermore I show similarities and differences between gestalt psychology and gestalt therapy on one hand and the theory of autopoiesis by Maturana and Varela on the other. After defining the goal of gestalt therapy as turning power relations into love relations, I explain the main concept in gestalt therapy "contact" and compare it with "structural coupling" in the theory of autopoiesis. At the end I quote part of a session with Fritz Perls, demonstrating some aspects of the therapeutic process and I show consequences of the theory for the relation between therapist and client.

Definition of Gestalt

GESTALT PSYCHOLOGISTS TRIED hard to define gestalt. There is no simple definition. Of course there are many definitions or descriptions of gestalt. The most influential ones are the definitions by the so called Berlin school of gestalt. Max Wertheimer, Kurt Koffka, Wolfgang Köhler, Kurt Lewin, Kurt Goldstein, Wolfgang Metzger, Adhemar Gelb belonged to this school. I don't want to show the differences to the other gestalt schools. And of course there are differences within the Berlin School of gestalt. The difficulties to define gestalt arise from our thinking habit, from our "habitus" (Bourdieu, 1980).

Max Wertheimer and Wolfgang Köhler saw our predominant thinking habit in "machine theory". You know that this thinking habit is still en vogue. The history of psychology shows that gestalt psychology nearly vanished after World War II when the machine theory of behaviorism became predominant. Nowadays machine theory is still predominant less in psychology but mainly in medical science, and of course in other sciences as well. I am convinced that nobody of us here is free of machine theory. Machine theory, said Wolfgang Köhler, is "*man's conception of nature since thousands of years*" (Köhler, 1947, p. 104). In parenthesis: this seems to be a wrong statement - Chinese thinking, the thinking of the American Indians, of Black Africa the thinking of our ancestors in neolitic age is very different from machine theory.

Machine theory says according to Wolfgang Metzger "... *that chaos can be prevented and order enforced, if proper controls are imposed upon from the outside* ..." (Metzger, 1976, p. 209). Let me emphasize this: Machine theory believes that the only way to prevent chaos and the only way to enforce order is by control *from the outside*. That means machine theory is a theory of control, the theory of control from outside. It is obvious that the fear of chaos, the anxiety of a nature, which is seen hostile, dictates machine theory/control theory. The contrary of machine theory is Gestalt theory. Max Wertheimer says that gestalt is something which is "*determined from within*" (Wertheimer, 1925, p. 7).

Wolfgang Metzger, who was Wertheimer's assistant in Frankfurt and Berlin calls the "law of natural order" the core ("Herzstück") of gestalt psychology. Machine theory is the law of natural disorder. The law of natural disorder says: "*Natural processes end in chaos if there are no forces from outside which control the process. Either control from outside or chaos*". As you know this is what the early thermodynamics said. But: "The law of natural order" says: "*There are natural processes, which realize order by themselves from within, without control from outside. Order arises by itself, is maintained by itself, changes by itself and recovers by itself (for example: healing)!*" (Metzger, 1976, p. 209) Wolfgang Metzger concludes: "*There is - besides the fact that there is order guided from the outside, nobody denies it - there is also a natural, inner, pertinent order, that is an order which is not forced upon but forms itself in freedom.*" (ibid.)

Metzger's law of natural order obviously has something to do with what we call self-organization nowadays. Most of the gestalt psychologists were mostly engaged in studying perception, as you certainly know, how we see, hear, feel etc. how objects outside form a gestalt. The so called Gestalt Laws describe that.

Kurt Goldstein was one of the gestalt theoreticians, who regarded a human being as a gestalt. Everyone of us is a gestalt. Goldstein was a psychiatrist, and a professor of medicine in Frankfurt. Fritz Perls, the founder of gestalt therapy, became his assistant.

Goldstein spoke of "organismic self-regulation". That is what he called gestalt. Goldstein, Köhler and Koffka also defined "life" as "spontaneous self-regulation" in their controversy with Driesch's neovitalism (Herrmann, 1976). Just to mention it: Max Wertheimer had an influence on Fritz Perls also by his theory on productive thinking, and Kurt Lewin influenced Perls by his field theory.

I am convinced that Fritz Perls was right in calling his therapy form gestalt therapy, because at his time the only formulation of the principle of self-organization was to be found in gestalt psychology, that is in the "law of natural order" or to take Goldstein's expression again in "organismic self-regulation". Perls was well aware that gestalt therapy was different from all other therapies. The main difference for him was: that in gestalt therapy human beings are seen as "*self-regulating and self-controlling organisms*" and that they are not seen as determined by the "*madness of control intervening from outside*" (Perls, 1981, p. 9).

One of the definitions of the neurotic by Perls says: the neurotic is somebody who "*uses all his energy manipulating others controlling others instead of growing by himself, by organismic self-regulation*". Adequate organismic self-regulation instead of letting oneself be controlled from outside, controlled by "*internalized shoulds, and instead of trying to manipulate or to control others - this is the goal of gestalt therapy*", as Fritz Perls saw it. (Perls, 1981).

Gestalt Theory and the Theory of Autopoiesis

In the following part I want to show similarities and differences between gestalt psychology and gestalt therapy on one hand and the theory of autopoiesis on the other. Max Wertheimer defined gestalt as "determined from within" and not determined from outside. This simply means: gestalten are autonomous and not heteronomous. Francisco Varela called his important contribution to the theory of autopoiesis "principles of biological autonomy" (Varela, 1979). Living systems are autonomous and they are not heteronomous. Autonomy does not mean that living systems are not determined, they are determined from within, therefore Maturana (1982) calls such systems "structure determined systems". Here gestalt theory and theory of autopoiesis are identical. The consequences of this autonomy thesis are enormous. They challenge our thinking very radically.

1.
Autonomous systems are operationally closed systems, this is the "closure thesis" by Varela. In cybernetics Ashby was the first who explored systems without input. Systems without input are cybernetic systems, where the output of the system becomes the input of the system. They are closed networks. The symbol of the closed network is "the snake eating its own tail" as von Foerster (1981) puts it.

Living systems are operationally closed systems. They are energetically open but they operate as closed networks. Structure determined systems are obviously influenced from outside, "perturbed", but what happens to the influence is "determined from within" the system and not from the outside. The effect of the influence from outside is determined from within, is determined by the structure of the system.

2.
I guess the most important consequences of the closure thesis - or of structure determinism - is, that we as living systems cannot perceive, represent, reflect the world outside. Everything from outside is determined from within - we construct the world outside, we don't represent reality or we don't reflect reality, we have to construct it. There is no objective knowledge. This is the construction thesis.

Ernst von Glasersfeld has shown that constructivism has a tradition in western philosophy (von Glasersfeld, 1985). The Greek philosopher Sextus Empiricus pointed to the problem that the perceiver cannot compare his perception with reality, he can only compare his perception with his perception - the reality stays outside. This of course brings us to the question how living systems as structure determined, operationally closed systems, constructing *a* not *the* reality, can survive.

The gestalt psychologist struggled with the question of constructivism and perception. They talked of the "erlebnis-jenseitige Welt", the world beyond experience, and the world of "immediate phenomenal reality". Wolfgang Metzger said: we cannot leave the world of perception: "*Wir können aus unserer Wahrnehmungswelt nicht heraus*" (Metzger, 1940).

Wertheimer and Köhler tried to solve this problem by their thesis of isomorphism between brain and reality.

As far as I can see there are three main ways of handling the problem of the relation between living systems and reality. This first is: the living system represents the reality or reflects the reality. This is the dominant view. The second way is: There is an isomorphism between living systems - especially the brain - and reality. This was the thesis of Köhler and Wertheimer and nowadays this is the thesis of Pribram and Bohm in their holonomy - or holography theory (Pribram, 1975; Bohm, 1980). The third way to handle this problem is the thesis of constructivism. The living system constructs *a* and not *the* reality. The gestalt psychologists, though recognizing constructions, nevertheless believe that the isomorphism between brain and reality warrants the correctness of our constructions of the world. Certainly Piaget is also a constructivist, but he also somehow believes in correct construction and so do the Russian psychologists like Leont'ev, who believes that "gegenständliche Tätigkeit" - action with concrete objects - provides feed-back about the real reality (Leont'ev, 1977). Constructivism is radical in saying that we construct *a* world not *the* world.

Maturana is emphasizing again and again to put "objectivity in parenthesis":

> "*Our incapacity of experientially distinguishing what we socially call illusion, hallucination or perception is constitutive in us as living systems*" (Maturana, 1987, p. 330).

> "*The assumption of objectivity, objectivity without parenthesis, entails the assumption that existence is independent of the observer, that there is an independent domain of existence, the* ***universum*** *that is the ultimate reference for the validation of any explanation. With objectivity without parenthesis things, entities exist with independence of the observer that distinguishes them and it is this independent existence of things (or entities, ideas) what specifies the truth ... He or she, who has access to reality is necessarily right in any dispute and those who do not have such access are necessarily wrong*". (Maturana, 1987, p. 332).

> "*Objectivity with parenthesis entails accepting that existence is brought forth by the distinction of the observer that there are as many domains of existence as kinds of distinctions the observer performs: Objectivity in parenthesis entails the multiversa, entails that existence is constitutively dependent on the observer and that there are as many domains of truth as domains of existence, she or he brings forth in her or his distinctions*". (Maturana, 1987, p. 332).

> "*Disagreements between observers when they arise not from trivial logical mistakes within the same versum but from the observers standing in different versa, will have to be solved not by claiming a privileged access to an independent reality but*

> *through the generation of a common versum through co-existence in mutual acceptance*". (Maturana, 1987, p. 332).

Perls wrote:

> "*There is no reality per se for human beings. Reality is something different for each individual, each group, each culture. The reality depends upon a reality of our interests, the inner reality not the outer reality ... I personally believe that objectivity does not exist. Objectivity of science is also just a matter of mutual agreement*" (Perls, 1969a, p. 13).

Assertions like this have consequences for therapy, I will talk about that later. First I will mention some other consequences of structure determined systems or systems determined from within.

3.
The only way how structure determined systems can exist is by self-referentiality. Self-referentiality says that every state in the system is determined only by states within the system. That is to say the system operates in a circular way. Russell and Whitehead prohibited self-referentiality in their Principia Mathematica in order to avoid paradoxes. One famous self-referential sentence is: "this statement is false". The taboo on self-referentiality is widespread. But Spencer Brown, a student of Russell, in 1967 went to Russell and showed him his calculus "laws of form" with the proof, that the so called theory of types with the taboo of self-referentiality was unnecessary by extending the valid arguments not only: true, false and meaningless but also imaginary - like imaginary numbers (Spencer Brown, 1979). This is the self-referentiality thesis.

Heinz von Foerster (1984) had shown that only by self-referentiality, by recursive operations we learn to "perceive" objects. One of his famous articles is called: "objects - tokens for eigenbehaviors"[7]. Values which are the result of recursive operations are called "eigenvalues" as you certainly know.

---

[7]see von Foerster, 1981.

Circularity makes the conception of causality - seen as linear causality - to a mis-conception. Fritz Perls said: "*It is much better to give up the causal explanation of events and to confine to descriptions instead - to ask how? instead of why?*" (Perls, 1969, p. 32). This is an important principle of gestalt therapy and it is of course in contrast to Freudian analysis. For Perls there are always mutual relations and no causal relations.

4.
Another consequence of autonomy or structure determinism is responsibility. Only autonomous beings can be responsible. Most people, as well as most of us don't like this radical responsibility. We are used to say: "I have to ...", "I must ...", "I cannot ...", "this and that brought me to do this and that ...", "my parents ..., my childhood made me do ...", "the circumstances are that ...", and so on and so on. One simple but impressive experiment in gestalt therapy is the following. Person A and person B sit in front of each other. A starts saying what he has to do like: "Every morning I have to get up to go to work. I have to eat. I must earn money. I must die". Then B formulates his obligations. Then both are asked to repeat their sentences substituting the "I must" by "I decide to" or "I decided to". So: "Every morning I decide to get up to go to work". "I decide to eat". "I decided to earn money". "I decide to die".

I believe it is really always a decision. I heard Virginia Satir say in her work: "You always have at least three choices and you always can decide". Fritz Perls was very precise in pointing to the responsibility of his clients again and again. You are responsible for your headache. You are responsible being bored etc. etc.

Francisco Varela (1979) pointed to the important distinction between prescription and proscription. A prescription is: You must do A and B and C everything else is prohibited. A proscription says: A and B and C are prohibited everything else does not matter. That is: there are always boundaries, limitations but within the limitations there is always an infinite number of possibilities. You know from mathematics that this is true, within the limitation there are always an infinite number of choices.

Maturana wrote: "*Everything that we do becomes part of the world that we live as we bring it forth as social entities in language. Human responsibility in the multiversa is total*" (1987, p. 378).

5.

This leads to another consequence of the theory of structure determined autonomous, operationally closed, self-referential, responsible systems, that is living systems. In this theory of living systems pathology cannot be formulated, that is living systems maintain in their praxis of living their organization by structural coupling in the medium and their adaptation or they disintegrate, that is they die. Structural coupling or adaptation is "*the relation of dynamic structural correspondence with the medium in which a unity conserves its class identity*", that is its organization, and "*which is entailed in its distinctions as it is brought forth by the observer in his or her praxis of living*" (Maturana, 1987, p. 338). This is the no-pathology thesis.

6.

In the definition of Maturana an observer is a "languaging living system". "*Everything said is said by an observer to another observer that could be him or herself*", is one of the famous quotations of Maturana. "*Whatever takes place in the praxis of living of the observer takes place as distinction in language through languaging and this is all that he or she can do as such*" (Maturana, 1987, p. 333). This is the thesis of human beings as languaging systems.

Language arises by structural coupling by coordination of actions in mutual perturbations. This is the linguistic domain. Language is the recursive mutual consensual coordination of actions of consensual coordinations of actions (or distinctions). That is the definition by Maturana (1987, p. 359). What I want to say is the following: By languaging - being in language - we become observers and self-observers by making distinctions and - by making distinctions and descriptions of objects and ideas and so on by languaging we operate on ourselves. Thus language is the "fall of man". "*A pathology is brought forth by somebody who defines something as a pathology - say a mental health problem and is accepted as an expert, whom others have given the social power to define something as a pathology*" (Mendez, Coddou, Maturana, 1988). Thus language brings forth pathology. But language on the other hand gives us the possibility to be observers, therefore it gives us freedom:

> "*The organism is free although his operation is deterministic when the organism generates consensual domains of second order*" (Maturana, 1982, p. 216).

He or she can generate as recursive observers of his or her circumstances consensual objects which are operationally independent from each other. Thus language is also the "salvation of man".

Fritz Perls was very explicit and expressive in his reservation against language. In his therapies he tried to bring his clients "to the senses", to bring them to the immediate direct experiences instead of abstractions. The slogan of Perls: "Loose your mind and come to your senses" was often enough misunderstood as rejection of the intellect or as rejection of the brain in favor of the "guts". Perls was very careful with language in his therapies and pointed to misuses of language again and again.

There is of course a difference when the client is saying: "I feel the tenseness in my hands and stomach", instead of saying: "I am tensing myself" (Perls, 1973). The influence of General Semantics certainly plays its part here. "The map is not the territory" is a slogan Perls liked to quote.

7.

Let me mention another important consequence of this theory of living systems or structure determined systems. Structure determined systems, living systems, human beings are determined from within by their structure but they cannot be predicted. The observer is not able to predict the course of the structural changes a living system will take. Maturana claims that the uncertainty principle of physics - the Heisenberg principle applies to living systems (Maturana, 1985b). This challenges psychology deeply. This is the non-predictability thesis.

Heinz von Foerster draws the distinction between trivial machines and non-trivial machines. Trivial machines - to quote Wolfgang Köhler again "man's conception of nature for thousands of years" - are simple input-output machines. Trivial machines are predictable, history independent, synthetically deterministic, analytically determined. Non-trivial machines are different. "*A non-trivial machine reduces to a trivial machine if it is insensitive to changes of internal states or if the internal states do not change*".

I quote this text by Heinz von Foerster because it shows how non-trivial machines can be trivialized. Non-trivial machines are synthetically deterministic, history dependent and analytically indeterminable, analytically unpredictable. Heinz von Foerster said: "*Our efforts to remove or suppress all uncertainty in our environment is quite understandable - we want trivial machines. However when we trivialize ourselves we shall soon not only be going blind, we shall also become blind to our blindness. Mutual trivialization reduces the number of choices*". So "the task at hand" is "de-trivialization". Heinz von Foerster's main ethical imperative is: "*Act always so as to increase the number of choices*" (von Foerster, 1984).

Fritz Perls was well aware of the unpredictability of living systems if they are not trivialized by socialization. Perls was a client of Wilhelm Reich and what Reich called character armour was seen by Perls as a basis for predictability: "*Society wants predictable characters*", trivialized machines. "*When we meet then I change and you change through the process of encountering each other, except the two have character. Once you have character you have developed a rigid system. Your behavior becomes petrified, predictable - and you loose your ability to cope freely with the world with all your resources. In our society we demand a person to have a character because then you are predictable*". (Perls, 1969, p. 7) But life is a process, an unpredictable process, things are events but we in our belief system believe that the future will be like the past.

8.

An important consequence of autonomy or structural determinism for gestalt therapy and the theory of autopoiesis is the boundary thesis. By being a closed network, that is by its circularity of operation - mind the snake eating its own tail - every living system builds its own boundary by its operation. In its praxis of living it makes the distinction between itself, the organism, and the environment. For Perls, Hefferline and Goodman (1951) the "organism-environment-field" is the subject of psychology. "*...psychology studies the operation of the contact boundary in the organism/environment field ... the contact boundary ... limits the organism, contains and protects it and* ***at the same time*** *it touches the environment ... it is essentially the organ of a particular relation of the organism and the environment ...*" (Perls, Hefferline and Goodman, 1951, p. 275). I will come back to the contact boundary in gestalt therapy later.

The Goal of Gestalt Therapy

Let me start with the quotation by Fritz Perls:

> "*So we come now to the most important interesting phenomenon in all pathology: self-regulation versus external regulation. The anarchy which is usually feared by the controllers is not an anarchy which is without meaning. On the contrary it means the organism is left alone to take care of itself without being meddled with from outside. And I believe that this is a great thing to understand that awareness per se - by and of itself - can be curative. Because with full awareness you become aware of this organismic self-regulation, you can let the organism take over without interfering, without interrupting we can rely on the wisdom of the organism. And the contrast to this is the whole pathology of self-manipulation, environmental control and so on that interferes with the subtle organismic self-control*" (Perls, 1969, pp. 17,18).

There are three possible relations between A and B:

1. No relation between A and B.
2. The relation is hierarchical. A controls B, A exerts power over B; or B controls, exerts power. A and B are heteronomous, controlled from outside.

3. The relation is non-hierarchical. A and B live in coexistence, in cooperation. The quality of this relation is very different to the power relationship. I call this relation "love", as Maturana did. A and B are autonomously determined from within, or structure determined.

Love, the acceptance of the other as an autonomous being is a biological phenomenon, as Maturana has shown, that is love is not something, one has to be socialized to. Love is not something extraordinary; it is nothing but spontaneous, dynamic, mutual, fitting together brought forth by recursive interaction conserving the individual organization and the mutual adaptation of living systems in their ontogeny (Maturana, 1985a, p. 129)

Love excludes power and power excludes love. Since love is a relation without power of A over B or B over A, love is a relation between autonomous unities. Let A be a human being like you and me then B can be another human being, your wife, your husband, a dog or any other living system, nature, a disease, an illness, a symptom, your own body... When you split yourself A may be you and B may be you. You may exert power upon yourself or you may love yourself.

The ultimate goal of gestalt therapy is to turn power relations into love relations, to accept oneself, to accept others, other human beings, to accept nature and so on as autonomous self-regulating, self-organizing entities. Let me put it in a different way. The ultimate goal of gestalt therapy is to give up our belief system that we can be and must be manipulated by others and ourselves and that we can and must manipulate others and ourselves. We have to give up "man's conception of nature for thousands of years", "machine theory" as Wolfgang Köhler called it.

That is not easy as you certainly know. I want to mention a very simple example. Sometimes I have difficulties to get to sleep. For me it is very difficult to really "fall asleep". I usually try hard to sleep, I want to make me sleep. I manipulate myself instead of letting go - that implies also to accept fully that I am not asleep at the moment. The famous gestalt paradox by Beisser says: "*Change happens when you become what you are, not when you try to become what you are not*" (Beisser, 1970). There is the well known cybernetic principle that no part of the system can control the system of which it is a part (McCulloch, 1969; von Foerster, 1981). I guess you have similar experiences, you know what I mean. One very famous gestalt technique is to let the client act the play between "top-dog" - usually the internalized "shoulds" of father and/or mother - and the "underdog" - often the rebellious or obstinate child - and to let the client integrate the two parts. Self-control, to be the master of oneself, to be in command of oneself is a well known ideal - a power relation. We control our pain by Aspirin, we treat our body like the body of somebody else, we don't know much about our body, we go to the physician and let our body treat as a machine. We try to control our body by various fitness-programs such as jogging. These are power relations. To love oneself is not easy. To accept yourself as you are, not to try to become somebody else, your ideal, (your "ideal self", as Perls put it) to be in solidarity with yourself, in contact with yourself, to have sympathy, compassion with yourself, to be one with yourself. Virginia Satir taught me, that if you want to love others you have to learn to love yourself first, to accept yourself, to be in solidarity with yourself, to be in contact with yourself, to have compassion with yourself, to be one with yourself.

Trying to control others in our everyday relationship we know very well and we are used to do that. Interestingly Maturana and Varela living or coming from Pinochet's Chile believe that power is not the problem but submission. Maturana says "*submission is the cause of power*" or "*who obeys grants power*". Do we really know, how we submit and when and where? (Krüll, Luhmann, Maturana 1987, p. 19).

In western society we conceive nature as something dangerous, an enemy we have to control. Our fear of nature is the cause of power relation to nature. Other cultures - American Indians for example, the Chinese - have a different relationship to nature. We also had it once what Berman calls 'participation'. Thc Hopi didn't see their rain dances to be the cause of rain, but they had to participate in the flow of nature. "Wu wei" is the expression in Chinese Taoism which is translated as "non-action". Needham (1977) says: "*wu wei*" means "*to respect the capacity of self-regulation*". The famous sentence from Tao-te-Ching is translated by Needham:

"Don't act (against nature) and there is nothing which is not in order". *"Don't push the river, it flows by itself"* is a very similar famous gestalt slogan.

In gestalt therapy theory the concept of "middle mode" is a rather central concept (Perls, Hefferline, Goodman, 1951, p. 430). In German and in English we have only active and passive verbs. Greek, however, has a regular middle mode. They define: *"Self is spontaneous middle in mode ...The spontaneous is both active and passive, both willing and done to ... a creative impartiality, a disinterest not in the sense of being not excited or not creative, for spontaneity is eminently these, but as the unity prior (and posterior) to activity and passivity, containing both"*. The spontaneous self is the "agent of growth".

In western societies to control nature we have to be "objective". *"Objectivity without parenthesis then is one of the most important basis of power. ... objectivity without parenthesis ... the claim for obedience..."* (Krüll, Luhmann, Maturana, 1987, pp. 16,17) I have already mentioned that. The German philosopher Gernot Böhme (1985) reminds us that objectivity is a means to protect ourselves, to keep distance. Compassion with nature, participation makes us suffer with something else.

One of the subgoals in gestalt therapy is contact. To be in contact with yourself, to be in contact with other human beings, to be in contact with "objects". In front of my studio window is a tree. Sometimes I don't see the tree I'm writing or thinking, my eyes are open. I look at the tree but I don't see it: No contact. Sometimes I look at the tree thinking: "Oh yes, the chestnut tree". Then I am in contact with the description, with the name, with the map not with the territory. Sometimes I am in contact with the identity of the chestnut tree. Here and now I can see it as an individual, unique, precious in its uniqueness. I and the chestnut tree are changing from moment to moment. This is contact at our boundaries.

And once or twice I felt there is no distinction between me and the chestnut tree - a mini-satori as Fritz Perls would say. I felt to be one with the chestnut tree, "non-two" as the Buddhists would say. And being one with the chestnut tree I made no distinctions, not only between me and the chestnut tree but also I made no distinctions between me/chestnut tree and the environment, all was one, non-two. This state is identical with the awareness of "emptiness" in the Buddhism of Nagarjuna. "*The originating dependently we call emptiness*" says Nagarjuna (Streng, 1967): nothing exists by itself, this is "originating dependently". We make the distinctions. We make the differences which make a difference, as Bateson would say. (Bateson, 1972).

When we become aware of the emptiness - the originating dependently - of everything *then suffering ceases*; this is the goal of Buddhism. In this moment there is no more attachment to anything, to no-thing, this is "nothingness", emptiness. Nagarjuna says there are two truths, the mundane truth and the ultimate truth. We can't live, talk without making distinctions, making differences, which make a difference. This is mundane truth. Ultimate truth is emptiness. We can become aware that everything is empty in everyday life. Turning power relations into love relations carried to its logical extreme is to be one with the world by ceasing to make distinctions. Then love of course stops to be a relation between A distinguished from B. That might not happen at all. But in therapy it may be enough to get a person to self-support instead of environmental support. "*Maturing*" - said Fritz Perls - "*is the transcendence from environmental support to self-support.*" (Perls, 1969). When you have self-support you don't need a therapist for environmental support; you don't need any person. Self-support is self-regulation, self-organization without control of ourselves or of others.

The Gestalt Therapy Process

It is the purpose of gestalt therapy to change power relations into love relations, I said. To change our belief system that we can be manipulated by others and that we can manipulate others. This is nothing else but realizing in every action that I and you are autonomous, self-referential, operationally closed, structure determined, always responsible, *a* not *the* reality constructing beings. Maybe this insight is sometimes a gestalt switch, a sudden insight. But why should we change our belief system into another one which is a belief system again, a theory constructed by us, neither the "reality" nor the "truth"?

Since we cannot claim by referring to correspondence to "reality" that one belief system is more "true" than the other, we have to rely our decision on other criteria, ethical or aesthetic ones. My interpretation is, that Maturana's claim for his construction is based on ethical principles, i.e. love (Krüll, Luhmann, Maturana, 1987). For Fritz Perls and Paul Goodman - the American anarchist - the goal of gestalt therapy is human growth, not so much healing of diseases; their decision is an ethical one too.

"Contact" is the central principle in gestalt therapy. The self is defined as the "system of contacts", as the "boundary at work", the boundary of the organism contacts the environment. Contact does not mean adjustment to "reality" as Freud formulated in his "reality principle", but "creative adjustment". Creative adjustment equals growth, and growth equals living. "*An organism preserves itself only by growing and it is only what continually assimilates novelty that can preserve itself and not degenerate... Contacting is ...the growing of the organism*". (Perls, Hefferline, Goodman, 1951, p. 427). "*Neurosis*", in contrast, "*is the fixation on the unchanging past, is 'contacting', the ever changing present environment as if it were the same as in the past, is not contacting the uniqueness of the momentary environment*" (p. 429). Thus neurotic adjustment, in contrast to creative adjustment, is inadequate "fixation on the unchanging past".

Habits are neurotic, the goal of gestalt therapy is "de-automatization" by awareness, as Laura Perls once said (personal communication). In the contact cycle there may be contacting disturbances: "Confluence" is the belief that organism and environment are identical, that there is no boundary between organism and environment (but there are "healthy" peak experiences of oneness for example in organism). "Projection" is the belief that something that belongs to the organism is in the environment. "Retroflection" is the belief that part of the organism is the environment of another part of the organism (for example auto-aggression). "Introjection" is the belief that something of the environment is in the organism (Perls, Hefferline, Goodman, 1951, p. 523). Submitting to the manipulation or the power of somebody else is introjecting the command of the other without "chewing" it, without "assimilating" it by determination from within, that is by structural determination. Obviously I cannot "contact" something which is inside me, there cannot be "creative adjustment". All "contacting disturbances" are "false" beliefs insofar as these beliefs do not fit to the boundary which a self-organizing, self-referential, operationally closed organism, just by its special organization, is continually preserving in its praxis of living in its environment. "Contacting" is very similar to what Maturana calls "structural coupling" of the living system to its medium (environment). The organismic structural coupling may be disturbed, because human beings are in language, they are observers, they construct a reality which may influence their organismic structural coupling. One can say: "*Disturbances in structural coupling arise by language and self-consciousness*" (see Maturana, 1987).

But we all have habits, an "action grammar", habits of thinking, perceiving, acting; we all are neurotics. And to change habits, that is the difficulty. Some say like the French sociologist Bourdieu that is impossible, for the praxis of living in this habits ("habitus") makes the class structure of our society. For example he has shown that our eating habits, our tastes differ from class to class (1980). Habits are "incorporated" in its true sense: They are in our body. Bateson believes that the change of habits, the gestalt switch is possible but not easy. He believes that you have to give up your "self", (in gestalt therapy terminology: your "personality"), that is your character. Can we live without habits, without character? In the terminology of Bateson this is learning III. Learning nonsense syllables is learning I, learning how to learn syllables is learning II, habit formation. Learning how we learn to learn nonsense syllables is learning how we form our habits, is learning III, so we may dispose of our habits (Bateson, 1972). Can we live without habits? But certainly it is not enough to change one habit for another habit, then we don't dispose of our habits.

Let me just mention the relationship between habit and distinction. Habitual acting is acting similarly in a similar situation. This presupposes that we conceive something as similar, and this presupposes that we make distinctions, we cut the flux of experiences in pieces, in objects to have similar and different objects. This is obviously "fixation on the unchanging past", neurosis.

We maybe cannot live without making distinctions but we can be aware of making distinctions - that is being aware of emptiness. So awareness - this kind of awareness - can be curative in itself as Perls said. That is to be aware of both truths of once mundane truth and ultimate truth.

I want to show how Perls was trying to provoke that kind of gestalt switch. It is a dialogue between the client Max and Fritz Perls:

1 M: I feel the chair, I feel heat, I feel the tenseness in my stomach and in my hands -

2 F: ***The*** *tenseness. Here we've got a noun. Now* ***the*** *tenseness is a noun. Now change the noun, the thing, into a verb.*

3 M: I am tense. My hands are tense.

4 F: Your hands are tense. They have nothing to do with you.

5 M: I am tense.

6 F: You are tense. How are you tense? What are you doing? You see the consistent tendency towards reification - always trying to make a thing out of a process. Life is ***process****: death is* ***thing****.*

7 M: I am tensing myself.

8 F: That's it. Look at the difference between the words "I am tensing myself" and "There's a tenseness here". When you say "I feel tenseness", you're irresponsible, you are not responsible for this, you are important and you can't do anything about it. The world should do something - give you aspirin or whatever it is. But when you say "I am tensing" you take responsibility, and we can see the first bits of excitement of life coming out. So stay with this sentence.

9 M: I am pushing down on the chair with my arms.

10 F: Are you sure? Do you experience this? ... Do it until you really feel ***you*** *are doing it, fully and a hundred percent responsible for what you are doing.*

11 M: I am holding my hands stiffly ... I am holding my whole body stiffly. My back is stiff - I am holding it stiffly.

12 F: Can you imagine what amount of energy is required to keep yourself so stiff, playing the corpse?

13 M: I can't go on because I am stiff.

14 F: Who's responsible for you being stiff?

15 M: ***I*** *am holding myself stiff. I have not relaxed myself.*

16 F: You have not relaxed yourself yet. See the split? "I am relaxing ***myself***".
17 M: But I'm not, as yet.
18 F: You feel you ought to relax.
19 M: I feel I can't go on until I relax.
20 F: You cannot go on. Who told you to go on?
21 M: I am telling myself that I want to go on.
22 F: "I am telling myself to go on". You're manipulating yourself. So you set up the king pin and then you try to knock it down. You make yourself stiff, and then you tell yourself to relax. You see all this energy that is wasting through this game?
23 M: I just relaxed myself.
24 F: You relaxed yourself?
25 M: I am more relaxed.
26 F: Did you do it, or did it happen?
27 M: It happened.
28 F: That's what I am talking about. Any deliberate change is doomed to failure. Change has to come by itself through organismic self-regulation. If you're hungry, you're hungry. If you eat when you're not hungry then you'll probably get a stomach ulcer ... I notice you're alive from the elbow down. You are like a dumpling and there's just a little bit sticking out - your hands. Otherwise you keep completely to yourself. Just become aware of this - how little expanding you are towards life. Now how do you feel about these remarks of mine?
29 M: I didn't like the word dumpling but I - it was true.
(Perls, 1969b, pp. 114-115).

I don't want to talk much about this example. You easily can find all the points I made: autonomy, responsibility and self support, awareness, manipulating oneself, organismic self-regulation and so on.

The Relation Client - Therapist

The decision for the principle of self-organization or for gestalt has consequences for the relation between client/patient and therapist. I gave six thesis to the students in clinical psychology when they were going into the institutions for their praxis phase, I want to repeat them here.

1. *Therapists who want to change clients by control from outside should stop to be therapists.*
   Clients are living systems and therefore autonomous. The therapist respects the ability for self-regulation. He/she is acting in the "middle mode" that is in the mode of "wu wei". All he or she can be is somebody like a midwife. The mother gives birth by self-regulation not the midwife. *Natura sanat, non medicus.*

2. *The client is always responsible for his/her actions* (thoughts, belief systems, problems) *and the therapist is always responsible for her/ his actions.* Autonomous systems are responsible. They are "response-able". All you can do is to elicit autonomous responses, which are determined from within.

3. *You cannot have no influence*
   When we are contacting I trigger a structural change in you, who is determined from within, and you trigger a structural change in me, who is also determined from within. This is contact between autonomous persons. Contact is always mutual. This relation is different from the power-submission - relation. I do not manipulate the client and I dont submit to the manipulations of the client.

4. *The special ability of the therapist is the "strange look"* (der fremde Blick", as Bert Brecht called it), *the "orthogonal procedure" as Maturana puts it* (Mendez, Coddou, Maturana, 1988).
   The definition of the problem by the client entails automatic behavior, he/she is not aware of, that is habits of perceiving, thinking, acting. By the "orthogonal procedure" the therapists keeps himself outside the belief system (the frame of reference)

of the client. Therefore he/she can make the client become aware of his/her tacit preconceptions and assumptions, his/her "habitus"; then he/she may dispose of the "habitus".

5. *It seems useful to look for the main problem of the client.*
   The question may go like this: How has the client constructed his/her world (reality) so that, as a result, he/she permanently does not fulfill his/her wishes and desires of main importance - or maybe there is only a pseudo-fulfillment?

6. *Growth is to increase the number of choices (von Foerster, 1981) by giving up habits.*
   The organism grows by "creative adjustment" that is by tasting and chewing what is novel in the environment . The process of therapy is the process of changing of the client by her/himself from the belief system that he/she is heteronomous to the belief system that he/she is autonomous. This is turning power relations into love relations. This can only be done by a love relation.

References

Bateson, G. (1972). *Steps to an ecology of mind.* New York: Dutton.
Beisser, A. (1970). The paradoxical theory of change. In: J. Fagan, I.L. Shepherd (eds.). *Gestalt therapy now* (pp. 77-80). Palo Alto: Science and Behavior Books.
Böhme, G. (1985). *Anthropologie in pragmatischer Hinsicht.* Darmstädter Vorlesungen. Frankfurt/M.: Suhrkamp, 1985.
Bohm, D. (1980). *Wholeness and the implicate order.* London, Boston and Henley: Routledge & Kegan Paul.
Bourdieu, P. (1980). *Le sens pratique.* Paris: Les Editions de Minuit.
Foerster, H. von (1984). Principles of self-organization - in a socio-managerial context. In: H. Ulrich, G.J.B. Probst (eds.). *Self-organization and management of social systems; insights, promises, doubts, and questions.* Berlin: Springer.
Foerster, H. von (1981). *Observing systems.* Seaside, Cal.: Intersystems Publ.

Glasersfeld, E. von (1985). Konstruktion der Wirklichkeit und der Begriff der Objektivität. In: H. Gumin, A. Mohler (Hg.). *Einführung in den Konstruktivismus*. München: Oldenbourg.

Herrmann, Th. (1976). Ganzheitspsychologie und Gestalttheorie. In: H. Balmer (Hg.). *Die Psychologie des 20. Jahrhunderts. I. Die europäische Tradition* (pp. 573-658). München: Kindler.

Köhler, W. (1947). *Gestalt psychology*. New York: Liveright.

Krüll, M., N. Luhmann, H.R. Maturana (1987). Grundkonzepte der Theorie autopoietischer Systeme. Neun Fragen an Niklas Luhmann und Humberto Maturana und ihre Antworten. *Ztschr. f. system. Therapie*, **5**,1, 5-25.

Leont'ev, A.N. (1977). *Tätigkeit, Bewusstsein, Persönlichkeit*. Stuttgart: Klett.

Maturana, H.R. (1982). *Erkennen: Die Organisation und Verkörperung von Wirklichkeit*. Braunschweig, Wiesbaden: Vieweg.

Maturana, H.R. (1985a). Reflexionen über Liebe. *Ztschr. f. system. Therapie*, **3**, 3, 129-131.

Maturana, H.R. (1985b). Biologie der Sozialität. *Delfin*, **V**, 6-14.

Maturana, H.R. (1987). The Biological foundation of self-consciousness and the physical domain of existence. In: E.R. Caianiello (ed.), *Physics of cognitive processes*. Singapore, New Jersey, Hong Kong: World Scientific.

McCulloch, W.S. (1969). *Embodiments of mind*. Cambridge, Mass.: M.I.T. Press.

Mendez, C.L., F. Coddou, H.R. Maturana (1988). The bringing forth of pathology. *The Irish Journal of Psychology*, **9**, 1, 144-172.

Metzger, W. (1975 (1940)). *Psychologie*. Darmstadt: Steinkopff.

Metzger, W. (1976). Gestalttheorie im Exil. In: H. Balmer (Hg.). *Die Psychologie des 20. Jahrhunderts. 1. Die europäische Tradition* (pp. 583-659). München: Kindler, 1976.

Needham, J. (1977). *Wisenschaftlicher Universalismus. Über Bedeutung und Besonderheit der chinesischen Wissenschaft*. Frankfurt/M.: Suhrkamp.

Perls, F. (1969a). *Gestalt therapy verbatim*. Lafayette: Real People Press.

Perls, F. (1969b). *In and out the garbage pail*. Lafayette: Real People Press.

Perls, F. (1973). *The gestalt approach and the eye witness to therapy.* Palo Alto: Science and Behavior Books.

Perls, F. (1981). *Gestalt-Wahrnehmung. Verworfenes und Wiedergefundenes aus meiner Mülltonne.* Frankfurt/M.: Verlag für humanistische Psychologie.

Perls, F., R.F. Hefferline, P. Goodman (1972 (1951)). *Gestalt Therapy.* Harmondsworth: Penguin Books.

Pribram, K.H. (1975). Toward a holonomic theory of perception. In: S. Ertel, L. Kemmler, M. Stadtler (Hg.). *Gestalttheorie in der modernen Psychologie. Wolfgang Metzger zum 75. Geburtstag* (161-184). Darmstadt: Steinkopff.

Spencer Brown, G. (1979 (1969)). *Laws of form.* New York: Dutton.

Streng, F.J. (1967). *Emptiness. A study in religious meaning.* Nashville, New York: Abingdon Press.

Varela, F.J. (1979). *Principles of biological autonomy.* New York: Elsevier.

Wertheimer, M. (1925). Über Gestalttheorie. Vortrag gehalten in der Kant-Gesellschaft Berlin, am 17. Dezember 1924. Erlangen: Verlag der philosophischen Akademie.

4

Toward a More Detailed Understanding of Self-Organizing Processes in Psychotherapy[8]

Henri Schneider

Abstract. The basic aim of the approach proposed in this chapter is to further the understanding of psychotherapy as a self-organizing process and to make the psychotherapeutic process amenable to research in the new cognitive science paradigm of connectionism. Two models which are intended as a theoretical framework for the detailed investigation of the client's change process will be sketched out.

IN THIS CHAPTER a framework for research on the psychotherapeutic process from the perspective of self-organizing processes will be presented. This theoretical framework should allow processes of change in psychotherapy to be described at a "detailed" level. By this I mean a level of description which enables the researcher to trace change processes in clinical material (Schneider & Wüthrich, 1988).

---

[8] I would like to thank Thomas Rothenfluh for his continuous support and for his comments on an earlier draft of this text.

The need for a more detailed theoretical framework arises from the fact that it has become possible to record psychotherapy sessions on audio or video tape, while the theoretical concepts used up to now evolved from the therapist's reconstructions of salient episodes from memory.

The starting point of the approach presented in this article is Prigogine's paradigm of *order through fluctuation* (Prigogine & Stengers, 1984). This paradigm seems to be well suited for the description of change processes in psychotherapy as it allows "microscopic" events (e.g., the effect of a single intervention) to be related to "macroscopic" entities (e.g. schemas). For the domain of psychotherapy, this paradigm is specified a an intermediate level of abstraction, using Piaget's model of *phenocopy* (Piaget, 1974a, 1975b).

The phenocopy model, which can be thought of as a kind of summary of Piaget's theory, is an attempt to describe evolution and cognitive development using the same theoretical language. In extremely simplified terms, and without entering into the discussion concerning evolution theory (cf. Bonner, 1982; Ho & Saunders, 1984), the basic idea of the phenocopy model is that a particular "structure" (genotype) produces an "active version" of itself (phenotype). During its development, the active version of this structure interacts with its environment, and this interaction in turn influences the final form of this phenotype, so that a certain phenotype may become quite different from prior ones generated by the same genotype. Now, the central point of the phenocopy model is that the genotype which has generated these different phenotypes may also undergo change by "re-constructing" the phenotype resulting from the interaction with the environment, i.e., by constructing a new pattern from which this new phenotype can be generated "directly". This reconstruction is effected by the genotype generating new variations (patterns for generating a certain phenotype) which are then selected by the "inner environment" of the actual phenotype. The selected variation will be the one which corresponds to, i.e., is able to generate, the "adapted" phenotype.

Due to the state of knowledge in theoretical biology and artificial intelligence at the time of his writing, Piaget was not able to work out the details of this model. The development of computational models from the perspective of self-organization, which has made major advances in recent years (cf. Pfeifer, Fogelman, Schreter & Bernold, 1989), will undoubtedly lead to a more differentiated language for the description of evolutionary and learning processes. If this language developing in the domain of connectionism is to be made fruitful for the understanding of psychotherapy (cf. Turkle, 1988), a basic "match" between the theory of self-organizing processes and therapeutic knowledge and intuition has to be effected. For this matching, Piaget's model of phenocopy is an extremely valuable conceptual foundation.

A word of caution may be in order at this point. Although in the long run the psychotherapist should be able to use the conceptual models proposed in this article to monitor the process of the client and to generate productive interventions, these models are not meant as guidelines for therapeutic action for immediate use. These models may well reveal new aspects of the client's change process to the psychotherapist, and in the discussion part I shall point out an emerging "invariant" that may ultimately prove to be a "therapeutic" model. The main emphasis of the framework outlined in this article, however, is on what is happening from moment to moment at a microscopic level. As we are just beginning to gather some impressions as to the nature of these processes, it will be some time before a coherent picture of change processes is available for therapeutic use at this new level of description!

In the remainder of this article, I shall first sketch out a basic framework for research on the psychotherapeutic process viewed as a self-organizing process. This account of how a structure changes in psychotherapy will be called the *selective activation model*[9]. In the second part of this article, I shall elaborate on a particular aspect of the selective activation model. Since I am taking up a model I had formulated in 1984 (Schneider, 1985), I shall continue to call it the *model of the negative class*.

[9]An earlier version of the selective activation model has been published in *Verhaltenstherapie und psychosoziale Praxis*, 1988, **20**, 24-38.

A Framework for the Description of Change in Psychotherapy: The Selective Activation Model

As mentioned above, the models I shall present are based on Prigogine's theory of *order through fluctuation* (Prigogine & Stengers, 1984). For the domain of psychotherapy, this approach is given concrete form by drawing on notions from Piaget's theory (1974a, 1975a, 1975b, 1977). The conceptualization of the psychotherapeutic process resulting from this perspective is closely related to Kohut's (1977, 1984) *Psychology of the Self* - in fact, I am constantly surprised at the extent to which this conceptualization has been "anticipated" by Kohut[10].

From this perspective, change is seen as a developmental process of a *cognitive system* (as Piaget uses the term[11], referring to Prigogine). Loosely speaking, cognitive systems correspond to what Mahoney (1985) calls *personal frameworks for meaning*: structures which lend significance to the environment, and which are responsible for the way in which the subject relates to significant others. These cognitive systems are thought of as "internalized object relationships" performing specific "functions", such as the "functions of the idealized father" (Kohut, 1977, p. 44) the activation of which leads to the sense of being encouraged or being told that

---

[10]The choice of the term "activation" is meant to reflect the interplay of the diverse frames of reference (order through fluctuation, connectionism, and psychology of the self).

[11]"*Ils [les équilibres cognitifs] sont ... plus voisins de ces états stationnaires, mais dynamiques, dont parle Prigogine, avec échanges capables de 'construire et maintenir un ordre fonctionnel et structural dans un système ouvert'...*" (Piaget, 1975a, p. 10). Piaget uses the term "cognitive system" in a large sense. The list he gives (1975a, pp. 70f.) encompasses among others descriptions, cognitive instruments (e.g. classifications), causal explanations, and more abstract structures such as groupings and groups. In a rather provisional manner, cognitive systems can be circumscribed as self-organizing cognitive entities at different levels of abstraction.

one has done something well[12]. This sense "produced" by the internal structure will be referred to as the feelings corresponding to the particular function.

Deficiency in a function leads to an inability to experience these feelings. A function has to be developed (during childhood or adolescence) with parents, siblings and peers in order to be able to be activated by everyday events later in life, which gives rise to an experience of the corresponding feelings. Deficient structures are conceived of as incomplete[13] structures having a tendency to complete themselves, in the sense of the Zeigarnik effect (cf. Piaget, 1975a, p. 138; Kohut, 1977, p. 217; Greenberg & Safran, 1987, p. 223).

The deficient structures undergoing change in the course of psychotherapy are thought to be specific for a particular client, i.e.,

---

[12]In the model presented in this article, the notions of structure and function are used in the sense of Kohut (1971, p. 50): "*The internal structure ... now performs the functions which the object used to perform for the child...*". *Structure* designates the constellation of a relationship, while *function* refers to the feelings produced by this ("internalized") structure. For a deeper understanding of the relationship between structure and function, it may be interesting to assume this relationship to be complementary in the sense of Bohr (1948, cited in Hermann, 1982; cf. Murdoch, 1987).

[13]An epistemological problem arises in relation to the notion of incompleteness (cf. Skarda & Freeman, 1987, p. 172): incompleteness can only be detected by an outside observer comparing a given pattern with a "complete" one. Nevertheless, I consider the notion of incompleteness of a psychic structure useful as a pragmatic shorthand to capture different intuitions:
- there is some kind of experience of this "incompleteness" by the client (the client may notice that he has difficulties in experiencing feelings he considers common in other people or he may experience some kind of diffuse unhappiness with certain aspects of his life);
- comparing this client with other clients, the therapist may identify an area of "deficit" in the client, i.e., an area where therapeutic work has to be done;
- the notion of incompleteness, together with the idea that a structural deficiency leads to attempts to fill in this gap, offers a new - and more "optimistic" - perspective of what has traditionally been called repetition compulsion.
In the theoretical framework presented in this article, an outside observer is not even needed for deciding when a structure has been completed (for this matter, this "observer" would have to be the therapist): "complete" structures simply do not show up any more in the therapeutic dialogue.

they result from the specific experiences (or specific lack of experiences) the client had as a child. This point would have to be elaborated; I hope the examples from psychotherapy I will use later on will help to elucidate what is meant by "specific". For the time being, I restrict myself to a quotation from Kohut (1977, pp. 216f.).

> *"On the basis of the preceding considerations we can now conclude that - as holds true with very few exceptions for all analyzable disorders, whether structural neuroses or self disturbances - the structure of Mr. X.'s psychopathology set the pattern for his analysis, that the specific course his analysis took and the specific remedial solution ultimately reached by it were predetermined. The essential transference (or the sequence of the essential transferences) is defined by preanalytically established internal factors in the analysand's personality structure, and the analyst's influence on the course of the analysis is therefore important only insofar as he - through interpretations made on the basis of correct or incorrect empathic closures - either promotes or impedes the patient's progress on his predetermined path."*

The basic idea of the selective activation model is the following. Cognitive structures are "looking for" an environment in which they can develop further. The client is enacting situations in which it is possible for him to experience the feelings corresponding to the deficient function. These enactments become more and more specific with respect to the underlying deficient structure. I would like to illustrate this model using two examples (Sandler, 1983; Le Guen, 1982). Sandler is reporting on a short period during a psychoanalysis of a 32-year-old woman, a physician. The incomplete structure being worked on in the therapeutic sequence presented by Sandler goes back to the patient's having been unable to work through with her mother her fear for her father and her pain at his death (the father had died when the patient was 7 years old). Le Guen's patient is in his late twenties and knows psychoanalysis, and the psychoanalytic literature, very well. The deficient structure in this example can be characterized as the patient's not feeling supported by his parents when he undertakes something new.

The Enactment of a Cognitive System: The Transition From its Inert to its Active Form

In order to be able to undergo change, cognitive systems enact themselves in the therapeutic situation or in everyday life[14]. Following Hofstadter[15] (1979, pp. 530ff.; 1985a, p. 30), this process can be conceived of as a transition of a cognitive system from its "inert" to its "active" form. The relation between the inert system and its enacted counterpart can be termed "self-referential". In using the notion of self-reference, I am referring to Hofstadter's treatment of this notion, and to his examples of direct and indirect self-reference[16] (1985b, pp. 58ff.).

The enactment is a slow process which may - as in the Sandler case - start with the client's evoking in the therapist the role of a significant other from childhood. When a situation is qualified as suitable in the sense that the client increasingly experiences feelings which correspond to the developing function (i.e., being able to deal with feelings of pain and mourning experienced toward the father), little by little an increasingly complete enactment takes place.

Sandler describes how her patient tried to evoke in her the role of her mother. "I noticed that I had difficulty ending the hour and that I repeatedly thought of Dr. M between hours" (p. 705). The

---

[14]Quite a few of these enactments actually take place in everyday life. But the only access to these enactments for the researcher would be through the client reporting on them in the therapeutic session. Furthermore, the model postulates that most of these enactments go unnoticed by the client! For these reasons, I restrict myself to enactments in the therapeutic situation itself.

[15]I think that Piaget's model of phenocopy and Hofstadter's *Central Dogmap* (Hofstadter, 1979, p. 533) are closely related not only in that they share a basic intuition of the workings of the mind, but also in that both authors use models from molecular biology to describe cognitive phenomena.

[16]To illustrate the notion of self-reference, Hofstadter uses - among many others - the examples of a couple talking about another couple (Matt and Libby, discussing the problems of Tammy and Bill...), or the two romances unfolding in parallel in the movie *The French Lieutenant's Woman*.

structure of the patient's relationship to her mother is shown by an incident the patient remembered when the analyst made clear to her how she was trying to actualize the image of her mother in the transference. The patient remembers that, on her fourth birthday, she was allowed to take a short train ride alone to her grandmother's house, and adds, "See, I knew that my mother would think about me the whole time and that the phone would ring as soon as I walked in Grandma's door." (p. 706)

Le Guen's article does not comment on the course of events through which the original relationship constellation is reproduced in the transference. His description begins at the point where the meaning of the structure thus created is already clearly recognizable.

The Formation of New Possibilities

In the active state of the cognitive system increasingly specific enactments are produced. This process is captured in the model of *aggregation by affective attraction* (Schneider, 1986, 1987). This model is based on the example of the construction of a termites' nest, used by Prigogine to illustrate the notion of order through fluctuation (Prigogine & Stengers, 1984, pp. 186f.). This model describes the first stage of the construction of a termites' nest, the construction of the base. At this stage, the termites transport lumps of earth and drop them in a random fashion. During transportation, these lumps of earth are impregnated with a hormone which then attracts other termites to the deposits. At one time or another, a slightly larger concentration of lumps of earth inevitably occurs at some point or other in the area. This small fluctuation is amplified by the slightly higher hormone concentration at the deposit, attracting more and more termites. With the increased density of termites in this region, the probability increases that they will drop their lumps of earth, leading in turn to a still higher concentration of the hormone. In this way, pillars are formed - a new structure comes into being.

The model of aggregation by affective attraction describes this process - the aggregation of elements to form a new structure - for the domain of psychotherapy. The model traces the process by which, little by little, a situation is brought about in which a self-referential experience may happen. *Self-referential experiences* are characterized by (1) the client experiencing the feelings associated with the deficient function and (2) the client being able work out in awareness the meaning of this experience.

The basic idea of the model of aggregation by affective attraction is that there are "elements" which are attracted and aggregated to form the situation in which a self-referential experience can take place. Elements can be characterized tentatively as aspects of the function under construction (cf. Kohut, 1971, pp. 50ff.)[17]. Situations in which the client experiences aspects of the function under construction without being able to work out the meaning of this experience are termed *self-referential situations*.

As in the termites' example, an initial fluctuation is amplified by elements being aggregated where a slightly bigger aggregation happens to take place, i.e., where several "right" experiences happen. A feeling resulting from such new experiences emerges in the client, which will then guide him in selecting further self-referential situations. (This feeling would be analogous to the higher concentration of the scent where pillars arise in the termite example.)

The attractiveness of an element is evaluated at an affective level, i.e., the client doesn't know from the beginning what the meaning of an element might be with respect to a self-referential experience in the making[18]. The attractiveness of an element for the

---

[17]The process of structure formation is called *transmuting internalization* by Kohut (1971, p. 49). Kohut links it to Freud's notion of "fractionized withdrawal of cathexes from objects": "... *the withdrawal of narcissistic cathexes takes place in a fractionated way if the child can experience disappointments with one idealized aspect or quality of the object after another; transmuting internalization is prevented, however, if, for example, the disappointment in the perfection of the object concerns the total object, ...*" (Kohut, 1971, p. 50).

[18]In the Sandler example, the client may feel some kind of relief from being able to talk to her therapist about the difficult experiences she had with her father; but even this "feeling" might not be accessible to the client at a conscious level during a large part of the process of aggregation of "elements", i.e., situations in which she has this experience of being able to talk to her "mother" about what is so painful for her. The therapist also may, at the beginning of this enactment, only be able to feel that something is happening which helps the client progress, without understanding yet the

cognitive system under construction results from its prospective contribution to the reduction of the disequilibrium (Piaget, 1974a, 1975b) or the lowering of the "computational temperature" (Hofstadter, 1984) resulting from the deficiency of the psychic structure under consideration. That is to say, elements are sought which on the one hand contribute to the reduction of the disequilibrium and, on the other hand, fit the developing structure. Those possibilities[19] - arising fluctuations - which reduce the disequilibrium relating to the deficient structure are the ones which are pursued further.

In every self-referential situation, the feeling resulting from the enacted function is amplified. This amplification, in turn, leads to the client's structures seeking more such experiences. In this manner, the particular new feeling resulting from experiencing the developing function in these self-referential situations is amplified, and at the same time, the constellations sought by the client become more and more specific with respect to the relationship structure underlying the particular deficiency.

Sandler reports that, after the three-week Christmas vacation, the patient was very concerned and anxious about the way the analyst's appearance had changed in the meantime. In the two sessions following the vacation, the patient now experiences more and more that it is possible to talk to her analyst (who represents the mother) about her grief at the loss of her father. I will take one brief episode from Sandler's description of the second hour following the three-week interruption in therapy.

After the patient has expressed her fear about the analyst's changed appearance, there is silence. Sandler describes how it occurs to her to connect the meaning which the patient attributes to her appearance with her (the analyst's) own worry about the changed appearance of a close friend who had been unsuccessfully operated upon for a malignant tumor.

---

meaning that this experience has for the client.

[19] "variations" in the phenocopy model (Piaget, 1975b, pp. 811ff.)

> *"I became sharply aware of the sadness and helplessness that rose up in me with the memory of the painful time with my friend. I was suddenly certain that Dr. M's silence represented her own attempt to keep back and to control the memory now rising up in her of an exceedingly painful and threatening event. I had the feeling that, in our mutual silence, she was including me in the process of her struggling to control herself. She was using me, so to speak, as a supportive auxiliary ego. When I realized this, I at once felt much more relaxed."* (Sandler, 1983, pp. 708ff.)

The patient seems to perceive this change in her analyst as a readiness to understand her.

> *"At that moment Dr. M decided to speak. To my surprise, she told me that she had suddenly had a vivid memory of a time when her father had come home from the hospital after a second operation. He had seemed to her a completely changed man. His hair, which had been black, had turned white (I think this might have been a screen memory of the bandage around his head). She continued, 'I know what scared me so much when you came back. You reminded me of my father! You look older, and your hair is grayer.'"* (Sandler, 1983, p. 709)

This sequence is characterized by the client's experiencing that it is possible to talk to her analyst about her fear - i.e., the client experiences an aspect of the developing function.

In the Le Guen case, the patient experiences the developing function by using ideas from a book written by his analyst for his own work. The feeling resulting from these activities constitutes a trace of the feeling of being supported by the father.

In these enactments, a constellation is being brought about in which feelings can be experienced which would correspond to the as yet incompletely established function. In the course of such enactments, which tend to become increasingly specific with respect to the developing structure, the client will be able to experience these feelings in a more and more distinct manner.

Self-Referential Experiences

Self-referential experiences - as is the case of self-referential situations in general - can take place in the therapeutic situation or in everyday life. One of the central ideas of the present model is that it takes time for a self-referential experience to become possible. The client has to have many experiences in self-referential situations before he is able to "understand" such an experience.

Self-referential experiences can be characterized in the following way. In a self-referential situation, the client experiences the function of the developing structure - in a situation in which the deficient structure is enacted - and is able to recognize the meaning of this experience[20]. An important aspect of the elaboration of the meaning of a self-referential experience may be that the client is able to establish a connection between the actual situation - in which he experiences the new function - and situations in childhood or adolescence in which he felt the lack of this function as especially painful. The structure of such a self-referential experience is very well captured by M.C. Escher's drawing "Print Gallery" (Hofstadter, 1979, p. 715). The client has a new experience and is also able to "see" the situation from the outside (cf. Kohut, 1977, p. 44; Wüthrich, 1987; Greenberg & Safran, 1987, pp. 224f.).

Again taking the Sandler case first: in the same therapy hour from which I selected a brief episode as an example of an experience in a self-referential situation, the meaning of this experience becomes clear. The patient told how, as a little girl, she had seen her father come back again from the hospital with a bandage around his head and had screamed loudly in fright. Her

---

[20]There is a problem with this formulation. It doesn't seem necessary for the formation of a new structure that the client does become aware of the meaning of a self-referential experience. At the same time, there seems to be a difference between "merely" experiencing the developing function in a self-referential situation and the ability to "grasp" somehow the significance of this experience. The important point may be that a new structure has come into existence of which the client *may* become aware. The "stabilization" of the developing structure taking place in self-referential experiences (see section on internalization below) may turn out to be more important than the explicit understanding of the meaning of this experience by the client.

father complained about the noise, so her mother took her out into the garden. The patient remembers that she then insisted on staying in the garden to play.

> "*She added that her mother must have thought her heartless and egotistical for having no other wish, at such a moment, than to play in the garden. I remarked that she might think herself egotistical now, for being concerned only with her own painful feelings, thinking that she ought to show concern for my feelings since, as she thought, something was seriously wrong with me. Dr. M tried to answer, but her voice broke and she started to cry. ... Then she said, 'I loved my father. He was the one who could make me laugh, who could play, who made me happy. My mother is all duty and routine. I always had the feeling that I had to look after her. I had so much fun with my father. When he got sick, all that was finished.'*" (Sandler, 1983, p. 710)

In this episode the patient experienced once again the function of a mother who listens, with whom she can share her grief. The therapist's interventions emphasize this aspect and, at the same time, help the patient work out the meaning of her experience. Sandler summarizes as follows these of the client's experiences which correspond to the deficient structure:

> "*We were able to understand how she did not dare allow herself to feel and express her deep sorrow at the loss of her father, for fear her mother would not be able to bear it.*" (Sandler, 1983, p. 710)

The best way to describe the self-referential experience of Le Guen's patient is to start from the interpretation of the analyst.

> "*You reproach Freud for not explaining to you how children are born.*" (Le Guen, 1982, p. 323)

Having heard this interpretation, the patient becomes agitated on the couch and falls into a long silence. Then he states that what the analyst just has said is overwhelming even though he doesn't know why. Two days later the patient reports that he has become uncomfortable and anxious. He attributes his state to the interpretation of the analyst. But he doesn't understand at all any more what the analyst wanted to tell him by his interpretation.

The patient complains that the analyst gave him an interpretation he couldn't understand - as he complains about his mother's not having given him suitable feet (the patient has flat feet like his mother). In the intervention of the analyst, the meaning of this (enacted) structure is touched upon (it is as if the analyst were to say: "You reproach me for not giving you what you need"), while at the level of the relationship this intervention represents an explanation, as the patient would have wanted to get it from his father.

On the basis of childhood episodes remembered by the patient during the sessions following this intervention, Le Guen elaborates the following constellation:

> "*... his mother had given what was best to his sister and she could have babies; furthermore, what his mother offers, his father takes back; his father explains nothing because he wants to keep it all to himself.*" (Le Guen, 1982, pp. 324f.)

One of these memories reported by the patient happened when he was 1½ or 2 years old, sitting on the potty in the bathroom. His father was shaving beside him. His father suddenly turned around, looked at him, grimaced, called for his mother and left. The patient never knew the reason for this departure, and he does not remember his mother appearing. But he thinks that the odor of his faeces must have disgusted his father. (The patient also had mentioned that he was extraordinarily ignorant of sexual matters. No one talked to him about these matters. His father in particular didn't give him real explanations, but was much more likely to make fun of him when he asked.)

No characterization of the patient's self-referential experience in this situation could be more to the point than Le Guen's:

"The indication that it was Freud who did not 'fertilize' him (in not revealing the mysteries of childbirth) mixed Freud's image with the image of a father grimacing at the sight of his stool; but my explanations allowed him some recourse, as when his father called for his mother. ... beyond my will, I led my patient to re-enact this trauma in the present. The trauma, unknown to me, caused me to readapt to its exigencies." (Le Guen, 1982, pp. 327f.)

"He now understood I had given him explanations his father had always refused him and that he had not found in Freud's works: I was his ally, the father who took away nothing, the mother who gave according to his needs." (Le Guen, 1982, p. 325)

How does the Enacted "Function" Become an Internal Function?

The course of change in a cognitive system described up to now may be summarized as follows. The cognitive system is enacted in different situations. Situations in which the client is able to experience feelings that correspond to the developing function are produced increasingly often, and these feelings can be experienced in an ever more marked manner. This process leads to the arrangement of a situation in which the function of the deficient structure can be experienced and the meaning of this experience can be understood. The question which has to be addressed now is: How does this enacted "function" become an internal function?

In terms of the phenocopy model, some "inner aspect" of self-referential situations which the client has lived - the feelings he has come to experience and some notion of the patterns he had arranged - would act as a "mould"[21] (Piaget, 1974a, p. 65; cf. 1975b, p. 812) in which the new internal patterns generated by the client are in turn formed (in an iterative manner; cf. Gernert, 1985). In a self-referential experience, the newly constructed (internal) structure

[21] In the introductory part of this article, this mould has been called "inner environment".

and the actual experience of the developing function in a new situation come to "fit". In this way, a self-referential experience would establish the certainty that the new structure corresponds to "reality", the new structure thus being stabilized. From this perspective, however, "reality" can only mean a situation brought about by the internal patterns of the client: the (new?) behavior of the significant other can be experienced as the newly established function only on the basis of these newly generated internal patterns. If the client has such "right" experiences (cf. Papert, 1980, p. 170), i.e., experiences which correspond to the variations generated, these will develop into a new, "stable" structure. In this frame of reference "internalization" takes on a new meaning: new patterns are generated internally , "suitable" variations will be amplified in self-referential situations and stabilized by self-referential experiences[22].

The client has developed new structures with which he arranges his relationships. Through the new kind of experiences which he has using these new structures, the newly developed function can be activated at the time, and the corresponding feelings can be experienced. In addition, the client may now look out more and more for relationships in which it is possible for him to have experiences activating these functions. Besides this, the new internal structures seem to be able to produce the internal experience of the new function "autonomously" (from memories?).

Change in the Client's Structures - without a Therapist?

What about the work of the therapist? Since the selective activation model conceptualizes change as a self-organizing process, it might appear difficult to describe the activity of the therapist in this frame of reference, particularly as we are used to thinking of a therapist's

[22]This formulation reflects the model of selective stabilization proposed by Changeux to explain the process of learning at a neurobiological level (Changeux, Heidmann & Patte, 1984; cf. Changeux & Danchin, 1974). In order to underscore the step-by-step aspect of the emergence of a new structure - processes at a microscopic level leading to a new macroscopic structure -, and to establish a reference to connectionism, the present model has been termed "selective activation model".

interventions as an active ("technological") change of some structure of the client. From the perspective of self-organization, however, the question would have to be stated in the following way: How to conceive of the therapist as contributing to the client's change process although not interfering with the client's autonomy?

The core of the answer to this question is that the therapist supports the process of generating new variations, i.e. internal patterns, used by the client to establish self-referential situations. In the therapeutic session, old patterns are broken up and new patterns can be created. With these preliminary new patterns the client is able to perceive his environment in a more differentiated manner and is able to arrange more specific situations, in which he can experience "new" aspects of the developing function.

In the therapeutic session, structures (internal patterns) are dealt with as they have determined particular situations - the client reporting on an episode in which a particular structure has been active. The therapist first aims at making "visible" this structure (e.g., by asking questions). On the basis of his understanding of this situation, the therapist tries to support the client in generating new patterns by proposing alternative views, showing connections to other situations, pointing out differences, and so on. Adopting the frame of reference of the Copycat model (Hofstadter, 1984), the interventions of the therapist are viewed as "enzymes" ("codelets") breaking up old structures and enabling the structures to establish new connections. These enzymes may look simple; however, they may be a complex product of the therapist's knowledge (knowledge about the client, experiences with former clients, models of typical situations, etc.). The more "sophisticated" these enzymes[23] are - i.e., the more "powerful" the heuristics they are based on (cf. Lenat, 1983) - the better they are able to contribute to generating new patterns which lead to ever more specific enactments.

---

[23] An example of what is meant by such enzymes might be an intervention which makes it possible for the client to experience the actual therapeutic situation. This intervention differentiates the actual situation from analogous childhood situations by commenting on the childhood experience - in the Sandler case:
"*I remarked that she might think herself egotistical now, for being concerned only with her own painful feelings, thinking that she ought to show concern for my feelings since, as she thought, something was seriously wrong with me.*" (Sandler, 1983, p. 710).

The overall state of a cognitive system may be captured by the metaphor of computational temperature (Hofstadter, 1984). This temperature depends on how "happy" (i.e., how appropriate for producing experiences corresponding to the developing function) the structures (internal patterns) are. The happier the structures, the lower the temperature, i.e., the less change will be necessary. The unhappier the structures, the more likely it is that enzymes are emitted to break up existing structures and permit the creation of new structures. Higher temperatures mean more "chaos" out of which new structures can emerge.

The temperature of a cognitive system can rise for different reasons, e.g. through disappointing experiences of the client or through (e.g. confronting) interventions of the therapist. By raising the temperature, i.e., by producing more "disorder", the emergence of new structures can be made possible in an unspecific way. However, the therapist has to take care not to endanger the "autopoiesis" of the cognitive system (cf. Brocher & Sies, 1986), i.e., the "coherence" of the client.

The Model of the Negative Class

Interventions which, on the one hand, reveal the theme behind the new experience and, on the other, capture the negative aspect of previous experiences, seem to be typical for situations in which the client has a self-referential experience[24]. This kind of intervention enables the patient to experience the current situation more fully (in the Sandler case: that the analyst is listening to the patient) by simultaneously highlighting the experience of the newly developed function (the patient being able to deal with her oppressive feelings) and separating it from experiences in childhood situations (that her mother didn't respond to her emotionally, being too much preoccupied with her own feelings). Only at this moment does the client seem to be able to recognize the negative aspect of previous experiences, i.e., to "conceptualize" the perturbation (that her

[24] As in Sandler's intervention (see the previous footnote).

mother couldn't stand her talking about her own sorrow). And only at this moment do such feelings, which arise from the fact that a function is missing, become comprehensible as an "object" (i.e., as an entity which can be referred to by the client and the therapist). The model of the negative class focusses on this aspect of the client's change process.

A short sequence on a videotape taken from the project of Christiane Schmid-Schönbein and Thomas Thiel (Schmid-Schönbein, Thiel & Rauh, 1983; Schmid-Schönbein, 1985) played an important role in the development of the model of the negative class. This videotape shows Wolfgang, a 4-year-old boy, trying to fit together the pieces of a "circular puzzle". In their project, aimed at investigating Piaget's equilibration theory, Schmid-Schönbein and Thiel conducted a longitudinal study on children between 3 and 5 years. At weekly intervals, children played with different materials, e.g. a "circular puzzle". Up to forty sessions by child have been videotaped.

The circular puzzle consists of four concentric rings arranged around a round piece in the middle and surrounded by a fixed frame of about 30 cm x 30 cm. The four rings are partitioned into pieces of equal size per ring, these being different however from ring to ring regarding their size and their curvature. Supposedly, a given piece can only be placed into the puzzle as a part of only one of the four rings, i.e., one of the four (sub-)classes. In fact, this holds only in principle, as we will see that in the situation taken as an example, Wolfgang lacks a piece because he has managed to misplace it earlier on.

In the case of Wolfgang, the "negative class" consists of pieces that don't fit into the puzzle at a certain place. As pointed out already, the task is very difficult for Wolfgang, after having put the piece he would need now in a wrong place. Encouraged by questions from the experimenter, Wolfgang starts to stack up equal pieces that turned out not to fit (previously, he had put them back anywhere on the table, with the result that he tried the same pieces again and again). Having made sure that none of the remaining pieces fit into the gap, he focuses his attention on the puzzle to look for a possible error.

This process can be described as an amplification of a perturbation into what will - at the next stage - constitute a subsystem with a "negative feeling tone". By this amplification of a fluctuation, a new class of objects (in the case of Wolfgang: pieces that don't fit) comes into existence in the sense that they not only are experienced as hindrances on the way toward a desired affective state (finishing the puzzle), but that they are conceptualized and can be referred to[25]. By this amplification of the perturbation, in turn, the overall system of the next level is activated. The "class" of pieces that don't fit acts as a perturbation and rouses Wolfgang's interest in the more encompassing system of which the negative class will be a part. This activation expresses itself in a heightened openness to new perceptions.

I would now like to apply this preliminary formulation to a vignette taken from a psychoanalytic case report (Goldberg, 1979, pp. 138f.):

> "*The narcissistic transference, which now appeared to be well established as a twinship transference, provided the patient with a relative sense of calm and harmony. This was shattered, however, when the analyst, in one of her interpretations, referred to Mr. M.'s grandmother as if she were still alive. The patient was very much attached to his grandmother; she was the only person he felt was truly invested in him. ... That the analyst did not remember that the grandmother was dead was unforgivable. In the past, whenever the analyst recalled a particular detail of his past, Mr. M. felt 'held' by her. Now he felt as if she had dropped him. The experience created a disruption in the self-object transference which resulted in a sense of confusion rather than in the patient's withdrawal into a haughty silence [as in the past]. ...*

---

[25]In terms of the approach of Yates & Kugler (1984; cf. Kugler & Turvey, 1987, pp. 64ff.), the process of amplification of a fluctuation by aggregating pieces which don't fit leads to the emergence of a new singularity which then takes on meaning for the subject. By this process "... *the sign and its significance are created (self-organized) simultaneously and jointly*" (Yates & Kugler, 1984, p. 59).

The self-object transference here was re-established (and an aspect of working through occurred) when the analyst interpreted the meaning which her forgetting the grandmother's death had for Mr. M. She said that forgetting something important to him was as if she had forgotten him or not cared about him. He felt carelessly dropped, similar to experiences he had as a youngster when he enthusiastically described something to his father, who either did not respond or changed the subject."

The cognitive system enacted in this episode is described in terms of a narcissistic (or self-object) transference. The patient's feeling of being accepted and loved is dependent on an object (in this case: the analyst) "holding" him. This is the feeling corresponding to the developing function. The above episode constitutes a self-referential experience for the client: He experiences the function under construction, and he seems to be able to grasp the meaning of this experience worked out in the interpretation of the analyst.

In terms of the model of the negative class, the change taking place through this self-referential experience would be interpreted as a construction of a class B encompassing the two subclasses A and A' (cf. Piaget, 1974b, pp. 166f.). A would correspond to a state which might be circumscribed by: "the psychoanalyst reacts positively to me, holds me, likes me...". A' might be paraphrased by: "the psychoanalyst doesn't like me", or "has abandoned me". This initial state of the cognitive system would be characterized by the patient's being in state A or A' depending on whether he feels accepted or not by his analyst at this very moment. The structure of the cognitive system of the next stage (i.e., when the encompassing class B can be conceptualized and experienced) would have to be expressed in the following terms: "The analyst likes me, even though from time to time she forgets things that are important to me."

Remember now Wolfgang's pile of wrong pieces. The basic idea of the model is that Wolfgang got the notion of "pieces that don't fit" (as opposed to a vaguer notion like "something doesn't work") from the amplifying of an initial perturbation, i.e., from the aggregation of wrong pieces. In the same way, the patient gets his notion of "being abandoned" (as opposed to his defending against such a feeling by a "withdrawal into a haughty silence") from the "piling up" of negative experiences effected in the interpretation of the analyst by the expression "similar to experiences he had ...".

By referring to a similar episode that the patient has experienced with his father, the analyst constructs a class of analogous experiences with a negative feeling tone. As the activation of the overall system brought about by the amplification of the perturbation increased Wolfgang's interest in the puzzle, this activation now leads to a greater openness to actually experiencing the developing function.

The updated version of the model of the negative class can now be formulated as follows. By stacking pieces which don't fit (or experiences of not having been "held"), a new structure emerges: the "class" of elements with a negative feeling tone. In the course of this process, the particular elements (pieces, states) become detached from the context of action and are identified ("impregnated") affectively: "as wrong as the other one" (or "similar to other experiences ...")[26]. This aggregation of affectively marked elements corresponds to the amplification of a fluctuation. By this process a new structure (a class of pieces which don't fit, or a class of specific experiences with a negative feeling tone) comes into being, and may now take on meaning. At the same time, this structure acts as a "catalyst" for the emergence of a new overall system - in the same way as the experiencing of the developing function makes possible the amplification of the perturbation.

What seems important to me is this process of the simultaneous emergence of a new structure and its negative "counterpart". The development of the new structure seems to go hand in hand with the possibility of conceptualizing (naming, specifying) the perturbation. To overstate it a little, this would mean that specifying the perturbation would be a prerequisite for the development of a new structure. (Some further evidence for this process may be found in Schmid-Schönbein (1985, pp. 176f.; cf. Forman (1981).) One might even think of a kind of transcatalytic association (Prigogine &

---

[26] This impregnation may be initiated by an interpretation of the "therapist": in fact, in the case of Wolfgang, the experimenter had asked, "How wrong is it?".

Stengers, 1984, p. 191) taking place between the emerging new structure and the negative class.

Discussion

In this paper, an attempt has been made to show how the paradigm of order through fluctuation may be applied to the psychotherapeutic process. The selective activation model and the model of the negative class are thought to make tangible "new" phenomena in the psychotherapeutic process and to make it possible to trace the change process of the client's structures at this detailed level of description. These models have been elaborated in a psychoanalytic "environment", but there is some evidence that they can be applied to other forms of psychotherapy as well (Wüthrich, 1987; cf. Greenberg & Safran, 1987, pp. 281ff.).

For further research on psychotherapy as a self-organizing process, it may also be useful to state the processes captured by the models presented in this paper in terms of attractors[27]. In this frame of reference, the developing function as well as the emerging negative class would correspond to attractors. The language of elements aggregating to form a new structure may turn out to be more helpful for the empirical tracing of these processes, whereas it may be more appropriate to use the language of geometrical dynamics (Abraham & Shaw, 1983) when the therapist's representation of these processes is concerned. Psychotherapy may then be viewed as a process of emergence of new attractors, leading to the formation of new structures.

---

[27] Prigogine & Stengers (1984, pp. 326ff.) establish the following relation between Waddington's approach based on catastrophe theory and the approach of order through fluctuation:
"*The concept of chreod is part of the qualitative description of embryological development Waddington proposed more than twenty years ago. ... C.H. Waddington's chreods are also a central reference in René Thom's biological thought. They could thus become a meeting point for two approaches: the one we are presenting, starting from local mechanisms and exploring the spectrum of collective behaviors they can generate; and Thom's, starting from global mathematical entities and connecting the qualitatively distinct forms and transformations they imply with the phenomenological description of morphogenesis.*"

As to the conceptual models proposed in this article, there are many questions which have to be answered by further theoretical and empirical work. One of these questions to explore is whether the working out of the meaning of a self-referential experience by the client is really necessary for change to occur (see also footnote 14). From the perspective of the selective activation model, this may not be the case. What seems to be important, however, is the activation of the new possibility (the client noticing again and again that his new way of going about something makes him or her feel good) and the stabilization effected by the realization that the experiences brought about by the newly generated internal patterns correspond to "reality", i.e., the realization by the client that the new way of seeing the world is "working". (The hypothesis that the elaboration of the meaning of his self-referential experiences by the client is a necessary prerequisite for change may be a remnant of the traditional ("macroscopic") psychoanalytic insight model...)

In scientific research, new phenomena to be investigated are "created" by the ongoing work and discussions in the field. New objects of research are emerging as "invariants" from the interactions of scientists. In the domain of self-organizing processes in psychotherapy, one of these phenomena seems to be situations which are characterized by Goudsmit (1989) as "merge between the object of discussion and the process of interaction itself", and which are called "self-referential experiences" in this article. These situations are considered to be important not only from the background of theories of self-organizing processes, but also in the context of therapeutic knowledge ("transference"; "transference interpretations" in psychoanalysis). The perspective of this article is that further investigation may lead to a new understanding of these situations at a more detailed level, which in turn may enable therapists to support their clients' change processes in a more differentiated manner.

Describing the psychotherapeutic process in terms of theories of self-organizing processes seems to be a promising agenda for research for many reasons. As the aim of this article was to present two conceptual models in detail, I would like to mention briefly only three of these reasons:

- the hope of deepening the understanding of change processes in psychotherapy by being able to relate moment-by-moment change events to change in the client's psychic structures;
- the prospect of describing cognitive processes and neurobiological processes using the same theoretical models, which may lead to new insights as to how these levels may be interrelated;
- the perspective of conducting cognitive science research on self-organizing processes in the domain of psychotherapy by referring to connectionist approaches.

References

Abraham, R.H., C.D. Shaw (1983). *Dynamics, the geometry of behavior. Part 1: Periodic behavior.* Santa Cruz, CA: Aerial Press.

Bohr, N. (1948). On the notion of causality and complementarity. *Dialectica*, **2**, 312-319 (cited in Hermann, 1982).

Bonner, J.T. (Ed.). (1982). *Evolution and development. Report on the Dahlem Workshop on Evolution and Development, Berlin 1981, May 10-15.* Berlin: Springer.

Brocher, T.H., C. Sies (1986). *Psychoanalyse und Neurobiologie. Zum Modell der Autopoiese als Regulationsprinzip.* Jahrbuch der Psychoanalyse, Band 10. Stuttgart-Bad Cannstatt: Frommann-Holzboog.

Changeux, J.-P., A. Danchin (1974). Apprendre par stabilisation sélective de synapses en cours de développement. In: E. Morin, M. Piattelli-Palmarini (eds.), *L'unité de l'homme. Vol. 2: Le cerveau humain* (pp. 58-88). Paris: Seuil.

Changeux, J.-P., T. Heidmann, P. Patte (1984). Learning by selection. In: P. Marler, H.S. Terrace (Eds.), *The biology of learning* (pp. 115-133). Berlin: Springer.

Forman, G.E. (1981). The power of negative thinking: Equilibration in the preschool. In: I.E. Sigel, D.M. Brodzinsky, R.M. Golinkoff (eds.), *New directions in Piagetian theory and practice* (pp. 345-351). Hillsdale, NJ: Erlbaum.

Gernert, D. (1985). Non-classical interactions in generalized neural nets. *Cognitive Systems*, **1**, 39-47.

Goldberg, A. (ed.). (1979). *The psychology of the self. A casebook.* New York: International Universities Press.

Goudsmit, A.L. (1989). Organizational closure, the process of psychotherapy and the psychologist's fallacy. In: G.J. Dalenoort (ed.), *The paradigm of self-organization.* New York: Gordon & Breach.

Greenberg, L.S., J.D. Safran (1987). *Emotion in psychotherapy. Affect, cognition, and the process of change.* New York: Guilford Press.

Guen, C. Le (1982). The trauma of interpretation as history repeating itself. *International Journal of Psycho-Analysis*, **63**, 321-330.

Hermann, P. (1982). Freud and the principle of complementarity (Letter to the Editor). *International Review of Psycho-Analysis*, **9**, 488-489.

Ho, M., P.T. Saunders (1984) *Beyond Neo-Darwinism. An introduction to the new evolutionary paradigm.* London: Academic Press.

Hofstadter, D.R. (1979). *Gödel, Escher, Bach: an eternal golden braid.* New York: Basic Books.

Hofstadter, D.R. (1984). *The COPYCAT Project: An experiment in nondeterminism and creative analogies.* MIT A.I. Memo 755.

Hofstadter, D.R. (1985a). Self-referential sentences: A follow-up. In: D.R. Hofstadter, *Metamagical themas: Questing for the essence of mind and pattern* (pp. 25-48). New York: Basic Books.

Hofstadter, D.R. (1985b). On viral sentences and self-replicating structures. In: D.R. Hofstadter, *Metamagical themas: Questing for the essence of mind and pattern* (pp. 49-69). New York: Basic Books.

Kohut, H. (1971). *The analysis of the self.* New York: International Universities Press.

Kohut, H. (1977). *The restoration of the self.* New York: International Universities Press.

Kohut, H. (1984). *How does analysis cure?* Chicago: University of Chicago Press.

Kugler, P.N., M.T. Turvey (1987). *Information, natural law, and the self-assembly of rhythmic movement.* Hillsdale, N.J.: Erlbaum.

Lenat, D.B. (1983). The role of heuristics in learning by discovery: Three case studies. In: R.S. Michalski, J.G. Carbonell, T.M.

Mitchell (eds.), *Machine learning: An artificial intelligence approach* (pp. 243-306). Palo Alto, CA: Tioga.

Mahoney, M.J. (1985). *Psychotherapy and human change processes.* In: M.J. Mahoney, A. Freeman (eds.), *Cognition and psychotherapy* (pp. 3-48). New York: Plenum.

Murdoch, D. (1987). *Niels Bohr's philosophy of physics.* Cambridge: Cambridge University Press.

Papert, S. (1980). *Mindstorms. Children, computers, and powerful ideas.* New York: Basic Books.

Pfeifer, R., F. Fogelman, Z. Schreter, T. Bernold (eds.) (1989). *Connectionism in perspective. Proceedings of an international conference, Zürich 1988, October 10-13.* New York: Elsevier North Holland.

Piaget, J. (1974a). *Adaptation vitale et psychologie de l'intelligence. Sélection organique et phénocopie.* Paris: Hermann.

Piaget, J. (1974b). *Recherches sur la contradiction. Vol. 2: Les relations entre affirmations et négations.* Etudes d'épistémologie génétique, Vol. 32. Paris: Presses Universitaires de France.

Piaget, J. (1975a). *L'équilibration des structures cognitives.* Etudes d'épistémologie génétique, Vol. 33. Paris: Presses Universitaires de France.

Piaget, J. (1975b). Phenocopy in biology and the psychological development of knowledge. In: H.E. Gruber, J.J. Vonèche (eds.), *The essential Piaget* (pp. 803-813). London: Routledge and Kegan Paul, 1977.

Piaget, J. (1977). *Recherches sur l'abstraction réfléchissante.* Etudes d'épistémologie génétique, Vols. 34 et 35. Paris: Presses Universitaires de France.

Prigogine, I., I. Stengers (1984). *Order out of chaos. Man's new dialogue with nature.* New York: Bantam.

Sandler, A.-M. (1983). Dialog ohne Worte. Nicht-verbale Aspekte der psychoanalytischen Interaktion. *Psyche*, **37**, 701-714.

Schmid-Schönbein, C. (1985). "He, sind ja beide gleich gross!" Eine prozessanalytische Rekonstruktion des Verständnisses von "gleich sein". In: T.B. Seiler, W. Wannenmacher (Hrsg.), *Begriffs- und Wortbedeutungsentwicklung* (pp. 167-189). Berlin: Springer.

Schmid-Schönbein, C., T. Thiel, H. Rauh (1983). *Der Aufbau kognitiver Strukturen als Äquilibrationsprozess (Abschlussbericht)*. Freie Universität Berlin, Institut für Psychologie.

Schneider, H. (1985). Modèles piagétiens comme point de départ d'une théorie du processus psychothérapeutique. *Schweizerische Zeitschrift für Psychologie und ihre Anwendungen*, *44*, 161-171.

Schneider, H. (1986). Piagets Begriff der majorierenden Äquilibration: Kognitive Entwicklung als selbstorganisierender Prozess. In: F. Stolz (Hrsg.), *Gleichgewichts- und Ungleichgewichtskonzepte in der Wissenschaft* (pp. 57-67). Zürcher Hochschulforum, Band 7. Zürich: Verlag der Fachvereine.

Schneider, H. (1987). *Order through fluctuation in psychotherapy: An attempt at concretization*. Paper presented at the 18th Annual Meeting of the Society for Psychotherapy Research, Ulm.

Schneider, H., U. Wüthrich (1988). *Den Prozess des Klienten nachzeichnen: die Entwicklung eines neuen Tools*. Paper presented at the 3rd Meeting of the PEP Project, Stuttgart.

Skarda, C.A., W.J. Freeman (1987). How brains make chaos in order to make sense of the world. *Behavioral and Brain Sciences*, **10**, 161-195.

Turkle, S. (1988). Artificial intelligence and psychoanalysis: A new alliance. *Daedalus*, **117** (1), 241-268.

Wüthrich, U. (1987). *Order through fluctuation in psychotherapy: Tracing a sequence in a case example*. Paper presented at the 18th Annual Meeting of the Society for Psychotherapy Research, Ulm.

Yates, F.E., P.N. Kugler (1984). Signs, singularities and significance: A physical model for semiotics. *Semiotica*, **52**, 49-77.

5

Anticipating Autopoiesis: Personal Construct Psychology and Self-Organizing Systems

Vincent Kenny

Abstract. George Kelly's theory of personal construct psychology is introduced in the context of comparisons with the radical constructivism theory of Ernst von Glasersfeld and the autopoietic theory of Humberto Maturana. Personal construct theory, although written in the decade up to 1955, anticipates in detail many of the epistemological and praxis issues currently concerning practitioners of psychotherapy. Following the comparative introduction, the formal aspects of Kelly's theory, namely, the Fundamental Postulate and elaborative Corollaries, are explicated within the framework provided by Maturana's theory. In this the chapter focusses upon the themes of change and stability within self-organizing systems. Finally, specific comments are addressed to the implementation of these theoretical prescriptions in the therapeutic relationship.

Introduction: Varieties of Constructivism

In using the term "constructivism", Kelly (1955) intended the dual meaning of "Constructive Vs Destructive" and also 'constructive' as in creating or inventing something new, some innovation or novelty. The central philosophy of the 'Psychology of Personal Constructs' is that of Constructive Alternativism which connotes the human capacity for audacious creativity and surprise. We do not have to be 'victims of our own biography' but rather we can 'transcend the obvious' by reaching beyond what appear to us to be inviolable 'objective truths'. That is, we can reach beyond our "psychological redundancies" (Kelly, 1977, p. 9) through processes of construing as opposed to 'representing' or 'simulating' the environment (Kelly, 1977, p. 4). The following quote illustrates the philosophy of alternativism on which Kelly's work is premised.

> "*What we think we know is anchored only in our own assumptions, not in the bed rock of truth itself, and that world we seek to understand remains always on the horizons of our thoughts. To grasp this principle fully is to concede that everything we believe to exist appears to us the way it does because of our present constructions of it. Thus even the most obvious things in this world are wide open to reconstruction in the future. This is what we mean by the expression constructive alternativism, the term with which we identify our philosophical position.*" (Kelly, 1977, p. 6).

He goes on to say that we will "*assume also that there is indeed a real world out there, one that is largely independent of our assumptions*".

In these quotes we see Kelly's emphasis on our own responsibility for personally inventing reality. He believes that there is no event which could be called "stark reality" because there is no event which we cannot reconstrue alternately. The central assumption is that we can invent something that is not already known or is not already in existence.

We can characterize the position in his early writings as that of 'trivial' (Vs radical) constructivism where he avows the existence of a real world which is there independently of any observer. His theory is that we construct a system or 'map' of the territory which we can continually improve by testing it out against the objective reality of the territory.

The objects of the real world can 'object' to our 'map' and cause us to change our system of axes, or our psychological space. Thus, the Kellian person exists in a dual reality, that of the real world and that which he makes of the real world (his constructions). Central to the characterization of a constructivist theory as 'trivial' is the notion that we can 'improve' or get better in our mapping of the world.

Ernst von Glasersfeld (1984) goes beyond Kelly's position into radical constructivism by stating that the most we can ever achieve is to discover what the world is not. Unlike Kelly he does not believe that our constructions of reality can 'match' (reflect) the independently existing reality, nor can we discover any "ultimate correspondence" between the two. This would be epistemological cheating; rather as if a blind man were to have occasional access to vision in order to check how well his touch sensations map was matching up with how things really looked. Instead, von Glasersfeld proposes that our constructions may 'fit' with reality in the way that a key may fit a lock. The fit of the key is defined in terms of its effectiveness in opening the lock and so fitting is a description of the key's capacity and does not refer to the lock. Furthermore, several different keys (constructions) may fit the same lock. This theory also encourages alternativism in that we should be flexible enough in our living to carry many 'alternative keys' to ensure our effectiveness when faced with life's locks. This position is radical because

> "... *it breaks with convention and develops a theory of knowledge in which knowledge does not reflect an 'objective' ontological reality, but exclusively an ordering and organization of a world constituted by our experience*" (von Glasersfeld, 1985).

Even in this radical position there is still this implacable 'real world' out there with which we continually collide with varying degrees of surprise and discomfort. Von Glasersfeld maintains that this 'real' world "*manifests itself exclusively there where our constructions break down*" (von Glasersfeld, 1984, p. 39). However, since we must construe these breakdowns in terms of our own (same) system of constructs which have already led to the failed structure (un-fitness), then we can never be in a position to know anything about the world that triggered the breakdown. We can only reconstrue our own system of ordering and its closed recursive processes.

Humberto Maturana develops this issue and criticizes the notion of radical constructivism as not going far enough. Maturana (1988) states that there is no objectively existing reality independent of some observer. Without the observer there is nothing (no-thing). Furthermore, due to the organizational closure of the nervous system, at the moment of experiencing we "*cannot experientially distinguish between what we call perception and hallucination*". The observer 'brings forth' his own reality by making his operations of distinction as the following quote makes clear.

> "*Therefore, we literally create the world in which we live by living it. If a distinction is not performed, the entity that this distinction would specify does not exist; when a distinction is performed, the created entity exists in the domain of the distinction only, regardless of how the distinction is performed. There is no other kind of existence for such an entity.*" (Maturana, 1978).

By putting (objectivity) in parentheses Maturana insists on reminding us that our reality is our own creation, completely dependent on our operations of distinction. From this position he criticizes radical constructivism for proposing the notion of 'fit' as the criterion of knowing where our theory breaks down, since this implies that we have access to an objective reality (independent of any observer) which can falsify our constructions.

> "*By putting objectivity in parenthesis we recognize that living together, that consensual operational coherences, that operations of distinction in language, constitute the generation and*

> *validation of all reality."* (Mendez, Coddou & Maturana, 1988).

Kelly in some of his later writings appears to be more radical and comes very close to Maturana's position where he emphasizes the personal nature of the construction of reality.

> *"Neither our constructs nor our construing systems come to us from nature, except, of course, from our own nature. It must be noted that this philosophical position of constructive alternativism has much more powerful epistemological implications than one might at first suppose. We cannot say that constructs are essences distilled by the mind out of available reality. They are imposed upon events, not abstracted from them. There is only one place they come from, that is from the person who is to use them. He devises them. Moreover, they do not stand for anything or represent anything, as a symbol, for example, is supposed to do".* (Kelly, 1970).

Here we see Kelly emphasizing the closure of the construct system, the ontology of the observer as Maturana calls it. However, it falls short of Maturana's position insofar as there are still 'events' separate from the observer upon which he places his constructions. For Maturana it is the construct distinctions which actually bring forth the events. The territory is the map. However, Kelly does repeatedly attempt to warn us of the danger of confusing objects and constructs, that the map is not the territory. It must be made clear at this point that a 'construct' is an individual act of discrimination and not the objects so discriminated, nor is it the verbal label we use to speak about such a sense-making discrimination.

> *"It must always be clear that the construct is a reference axis devised by a man for establishing a personal orientation toward the various events he encounters. It is not itself a category of events or even the focus of a class".* (Kelly, 1969, p. 10).

Therefore, in construing we are relating our 'selves' to the events at hand within the personal orientation domain generated by our construct (meaning) system. This relates to later comments below on the cognitive domain within which we are oriented, simultaneously, toward effective interactions in our medium, and to the conservation of our organization.

In discussing a related theme, that of 'representing' events as opposed to 'construing' them Kelly comes very close to one of the most radical aspects of Maturana's theorizing, i.e. where he abandons the notion of mental representations that correspond to events in the world. Since for Maturana the nervous system is organizationally closed, and he has no place for notions such as information being transmitted, or inputs arriving to the organism, or the independent existence of 'things' outside the observer, then he has no need to postulate mental models which try to accurately represent reality. Rather he substitutes his notion of 'structural coupling' in the place of 'representations'.

Kelly does not go so far but he does attempt to substitute his notion of construing for that of representation. Instead of attempting to simulate or symbolically reproduce events we must rather develop approaches to events which can help us to achieve a transcendent understanding of them.

> "*It is sometimes said that man is the only animal who can represent his environment. The implication is that this ability to simulate reality is the principal psychological feature of man, and, hence, something we all ought to cultivate. To be sure, the reproduction of reality is one of the post hoc tests of understanding - prediction, for example, is another. But the mere reproduction of reality adds precious little to one's understanding of it. Man can and does do better than that. This, of course, is something every artist knows. As for the artist, so for all of us; to construe the surrounding world is to visualize it in more than one dimension. And this is no less true for the psychological theorist. He, like the artist - and man - must transcend the obvious. A simulative theory is not enough; in fact it is no theory at all, only a technician's blueprint for the reproduction of something.*" (Kelly, 1977, pp. 3-4).

This clear rejection of 'representation' is an aspect of these theories most likely to shake our 'common sense' notion that we can and do have clear representations which correspond to our immediate environment. If we remove this common 'sense' then it seems to be very difficult to imagine what type of world we are left with. In fact, Ernst von Glasersfeld invites us to view our environment as a "black box" because we can never come to know it as if an objective reality.

> "*What we experiment with ... is, in the last analysis, never anything but the interrelation of our signals which we have come to consider input to, and output from, the black box of the "universe"; and what we modify or control by our activities are always, as William Powers (1973) has formulated it, our own perceptions, i.e., the signals we call sense data, and the ways in which we coordinate them. ... even the most spectacular successes we achieve in predicting and controlling our experience give us no logical ground whatsoever for the assumption that our constructs correspond to or reflect structures that "exist" prior to our coordinating activity.*" (von Glasersfeld, 1987, p. 108).

With these comments on self-coordination we may now proceed to the next section.

Self-Organizing Systems

The second law of thermodynamics states that order is unstable and that processes inevitably move towards an increase of chaos and breakdown. However, a number of theories have emerged in the recent past which argue that the opposite is also true in certain physical, chemical and biological phenomena where it is found that it is chaos which is unstable. There has been a shift from the previous concern with the itemization of static components to a process-oriented view, from 'being' to 'becoming'. In his attempts to propagate the constructivist process view of persons, Heinz von

Foerster suggests that we should no longer call ourselves 'human beings' but rather 'human becomings'. His work in the Biological Computer Laboratory (University of Illinois) emphasized the self-organizing features of living systems. Ilya Prigogine's work on dissipative structures in chemical systems led to the development of a new ordering principle which he called 'order through fluctuation'. Erich Jantsch's (1980) theorizing on self-organizing systems led him to attempt to integrate a variety of theories of self-regulation and self-organization within the framework of the phenomenon of dissipative self-organization. In particular he tried to unify within his paradigm Prigogine's theory of dissipative structures (order out of chaos), Maturana's concept of autopoiesis (self-production) and Eigen's (1971) theory of self-reproducing hypercycles.

Jantsch viewed autopoiesis as "*one of the ways in which the self-organization of non-equilibrium systems manifests itself*". (Jantsch, 1981, p. 66). He also states that

> "*Autopoiesis refers to the characteristic of living systems to continuously renew themselves and to regulate this process in such a way that the integrity of their structure is maintained. Whereas a machine is geared to the output of a specific product, a biological cell is primarily concerned with renewing itself.*" (Jantsch, 1980, p. 7).

These theories are unified by being oriented to generating models of living, interactive systems which are characterized by the features of autonomy (self-determination), self-organization, and self-renewal.

While the notion of autopoiesis was invented specifically for the context of the cellular domain and does not translate readily to the domains of the social or psychological, any system, biological or otherwise can be analyzed in the more generalizable terms of organization and structure. The structure of a system is defined as the concrete components and the actual relations that exist between them which realize or materialize the system as a particular composite unity characterized by its particular organization.

> "*In other words, the structure of a particular composite unity is the manner it is actually made by actual static or dynamic*

> *components and relations in a particular space, and a particular composite unity conserves its class identity only as long as its structure realizes in it the organization that defines its class identity.*" (Maturana, 1987, p. 335)

What then is the organization? This refers only to a subset of all the actual relations of structure. For the organization to retain its class identity invariant this specific subset of relations among components must be conserved.

> "*In other words, the organization of a composite unity is the configuration of static or dynamic relations between its components that specifies its class identity as a composite unity that can be distinguished as a simple unity of a particular kind.*" (Maturana, 1987, p. 334).

Therefore, all changes must occur through the structure of the system. The system is structure-determined. Its structure may change endlessly while keeping the organization invariant. The system lasts as long as its organization (class identity) is conserved. Living consists in the conservation of identity. This is realized in structures which change continuously. There are only two types of structural change, that in which the organization is maintained (changes of state) and that in which the organization is destroyed. These two categories of change have a correspondence to the emergence of psychological difficulties in personal construct psychology. We can define psychological distress in terms of (1) Structure, where a person repeatedly uses some construct system structure despite having experienced consistent invalidation of it, and (2) Organization, where the structural changes are so widespread throughout the interconnected system that the whole system disintegrates and the current organization (identity) is lost.

As we have seen, Maturana defines a living system as one which must conserve both its organization (identity) and its means of "fitting" with its environment (structural coupling). In Kelly's terms, the core constructs of the system are the self-maintenance processes (conserving our organization), while our subordinate system of constructs provides the instrumental channels whereby we directly relate ourselves to the world as embodied subjects. The whole personal construct system can therefore be seen as organizationally closed while being structurally open. It is in the latter that we find the way the person changes himself in order to conserve his stability (Keeney, 1983). Dis-order or dis-organization arises when either level of conservation is threatened. That is, dis-order of the relations of 'fitting', or dis-order of the relations of self-constitution. Since the system cannot make mistakes, what is usually called a 'symptom' is merely an alternative means of 'fitting' with the world, of relating oneself to one's circumstances, of carrying out an experiential experiment, or, in Maturana's terms, it is a way of structurally compensating for an environmental perturbation. The more the person does (practices) the 'symptom', the more it is recursively recycled into his core-role (identity). In this way the person cleaves or specifies a meaning-space within which he can effectively act to maintain his organization. (His organization specifies his cognitive domain). In all of this it is important to bear in mind that constructs are not something that one 'has' (like instruments or possessions) which one may 'apply to' external objects or events. Rather, you are your constructs. 'Being' is brought forth through enacting what is 'known'. The 'construer', the 'construing', the 'constructs', and the 'constructions' are all one indivisible wholeness.

In Maturana's view the circular organization of the living system is premised on the assumption that previous interactions will recur. Since the organization specifies the interactions that the system can engage in, this prediction of recurrent interactions is crucial since if they do not recur, the system will disintegrate. Maturana continues thus:

> "*Accordingly, the predictions implied in the organization of the living system are not predictions of particular events, but of classes of interactions. Every interaction is a particular interaction, but every prediction is a prediction of a class of interactions that is defined by those features of its elements that will allow the living system to retain its circular organization after the interaction, and thus, to interact again. This makes living systems inferential systems, and their domain of interactions a cognitive domain*" (Maturana, 1980, p. 10).

This is also something that Kelly certainly agrees with since at the heart of his theory is the notion that a construct system is an anticipatory system of predictions and inferences. Because the system of constructs is interrelated and many lines of inference are to be found throughout the network of constructs they can form the basis of consistent and effective anticipations. Constructs are both descriptive and prescriptive, i.e. they describe events and tell us what to do with them.

In discussing the other issue which Maturana deals with in the above quote, i.e. that what is predicted is a class of interactions, Kelly has this to say:

> "*A person anticipates events by construing their replications. Since events never repeat themselves, else they would lose their identity, one can look forward to them only by devising some construction which permits him to perceive two of them in a similar manner....Perhaps it is true that events, as most of us would like to believe, really do repeat aspects of previous occurrences. But unless one thinks he is precocious enough to have hit upon what those aspects will ultimately turn out to be, or holy enough to have had them revealed to him, he must modestly concede that the appearance of replication is a reflection of his own fallible construction of what is going on. Thus the recurrent themes that make life seem so full of meaning are the original symphonic compositions of a man bent on finding the present in his past, and the future in his present.*" (Kelly, 1970, pp. 11-12).

The emphasis on the anticipation of recurrences tends to make the system a conservative entity prone to 'self'-fulfilling prophecies and increasing rigidity in the repertoire of viable ideas. To steer us away from rigidifying formulae and to keep us in the spirit of constructive alternativism, Kelly polarizes a 'world of chaos' against a 'world of certainties' and warns us to beware of the obvious. While many people are tortured by the uncertainties of life, Kelly is suspicious of narrowing down our experience to things we can appear to know with assurance.

> *"But those things I once thought I knew for sure, those are what get me into hot water, time after time. They are a lot more troublesome than those things I have known all along that I didn't understand. Moreover, a world jam-packed with lead-pipe certainties, dictionary definitives, and doomsday finalities strikes me as a pretty gloomy place. How can there be any room in a world like that for such a nascent thing as life?*
>
> *I suppose it would have to be conceded that life in the opposite kind of world - a world of chaos - might seem pretty hopeless after a while. But, as between the two - a world without hopes or a world without doubts - I think for myself, I would prefer the world without hopes."* (Kelly, 1969, p. 51).

Kelly therefore encourages us to destabilize our construct systems and to become dislodged from our sense of self. The process whereby we should do this is spelt out in his five-phased Cycle of Experience which begins with Anticipation where we begin to project our selves forward into oncoming events, continues with personal Investment and experientially Encountering the predicted events, next the emergence of Confirmation or Dis-confirmation (of one's initial Anticipation), and finally the Constructive Revision of one's system. This leads one back anew into a fresh experiential cycle beginning with new anticipations and so on. As far as Kelly is concerned the cycle of experience is not complete unless it concludes in *"fresh hopes never before envisioned"* (Kelly, 1977, p. 9). The more we can invest ourselves in our anticipations and constructive revisions following falsification, the more vivid is our human experience. Kelly notes the following:

> *"But if he invests himself - the most intimate event of all - in the enterprise, the outcome, to the extent that it differs from his expectation or enlarges upon it, dislodges the man's construction of himself. In recognizing the inconsistency between his anticipation and the outcome, he concedes a discrepancy between what he was and what he is. A succession of such investments and dislodgements constitutes the human experience."* (Kelly, 1970, p. 18)

Symptoms emerge from uncompleted cycles of experience, where the client has become unable to fully engage in the cyclical flow. People become stuck at various parts of the cycle, for example by having such ambiguous anticipations that they cannot be fully specified in action; by being too afraid to risk themselves in commitment to personally invest themselves in the anticipatory prescriptions; by fearing encounter so much that when the event arrives they feel too threatened to fully 'indwell' it to use Polanyi's (1958) phrase; by having a system which refuses to construe any perturbation as a falsification; or by having a system which refuses to construe the outcomes of falsification, or, having done so, the revised construct "*is left to stand as an isolated axis of reference*" (Kelly, 1970, p. 19) and so remains largely ignored. The emergent novelty or change must be cycled into the construct system if any change in the system is to take place. Kelly observes that

> "*....A symptom was an issue one expresses through the act of being his present self, not a malignancy that fastens itself upon a man. What they experienced as symptoms were urgent questions, behaviourally expressed, which had somehow lost the threads that lead either to answers or to better questions. The symptom was even a fragment of proper human experiment - one designed in childhood, perhaps, and repeated again and again in later years. Yet the experiment could never be carried through to its conclusion because the generating hypotheses had lost their contexts or because the current outcomes had slipped out of focus.*" (Kelly, 1969, p. 69).

When we complete an experiential cycle we have revised our construction of events and also, perhaps more importantly, may have revised our perspective on the processes whereby we arrived at our new conclusions. In other words, at the outset of our next cycle we will have not only our new anticipation of events but also a new anticipation regarding the 'effectiveness of the experiential procedures' we used last time round. Our construing of the way we go through the cycle of experience is a powerful mechanism for changing the ways we allow ourselves to change. Much of therapy is directed at changing the ways the client changes himself. From this point of view, symptoms are reframed as the way the person changes himself in order to remain organizationally invariant. In 'choosing' to live in a world of no doubts a person may transform his experiential chaos into the order of symptoms. Another way to transform personal chaos and attempt to maintain one's organization is, paradoxically, to commit suicide. Taking affirmative action in the face of chaos is a way of validating what remains of one's core structure.

The Personal Construct System as a Self-Organizing System

What I hope to show here is that Kelly's views on personal construing are premised on treating the system as a closed network of self-production within a structure-determined framework. The corollaries of the formal theory variously emphasize the issues of conservation of organization in relation to possibilities for structural change. In what follows I group the various corollaries around these issues, although this is somewhat artificial, since they form a complex whole and so cut across simple change/invariance boundaries. It has already been implied that the notion of 'pathology' has no place within Kelly's theory. Each one of us is 'symptomatic' to the degree that we abandon our various cycles of experience part-way through, leaving unexplored questions, unexpressed commitments, unapproached encounters, unexperienced dislodgement, or unintegrated novelty scattered about our experiential space. In what follows, the role of the therapist as a co-experimenter with the client in activating, unfolding and elaborating the construing system must be kept in mind. Kelly characterized this role as that existing between a research student and his supervisor. The student is the expert in his own subject matter (in this case himself), and the supervisor is an expert in how to design viable experiential experiments so that the joint explorations may lead to even more constructive living questing.

We now turn to examine the formal aspects of Kelly's theory - the Fundamental Postulate and the eleven elaborative corollaries - in terms of Maturana's distinction of organization/structure.

The Fundamental Postulate states the central core of the theory of personal constructs. It says:

"A person's processes are psychologically channelized by the ways in which he anticipates events."

The philosophy of constructive alternativism is embedded in this statement along with much else. Kelly sees his work as a metatheory, i.e. a psychological theory about how we can create psychological theories, and the inherently idiosyncratic personal nature of such creations. Having specified that the person is a form of movement, what has to be explained, is the direction of the personal processes and not the transformation of static states into processes (e.g. 'motivation theories'). While Maturana explains the direction of drift in terms of the moment to moment co-ontogenic structural interactions of the system in its medium, (while conserving its organization and adaptation i.e. structural coupling), Kelly, for his part, accounts for direction in terms of ways of anticipating events.

> *"...neither past nor future events are themselves ever regarded as basic determinants of the course of human action - not even the events of childhood. But one's way of anticipating them, whether in the short range or in the long view - this is the basic theme in the human process of living."* (Kelly, 1970, p. 10).

One's ways of anticipating are of course determined by the repertoire of structure existing at any one moment. Earlier we have seen that Kelly and Maturana substituted their respective notions of 'construing' and 'structural coupling' in the place of 'mental representations'. In this passage they substitute the same concepts (anticipatory construing and structural coupling) as their explanatory principle for direction of drift. That is, the ways in which we structurally couple with/construe events gives the channelized direction of drift to our social-psychological processes.

Leaving this discussion of Kelly's Fundamental Postulate we will now briefly examine his eleven corollaries in terms of Structure and Organization, especially in relation to the issues of change/stability and the Cognitive Domain in Personal Construct Psychology.

Structure

Components: An important feature of the components (constructs) of the system is defined by the Dichotomy Corollary which states "*A person's construction system is composed of a finite number of dichotomous constructs.*" According to Kelly, thinking is essentially dichotomous. Anything which can be said has an implied contrast which may either be obvious or difficult to articulate. He wishes to emphasize the intimate interpenetration of opposites like Heraclitus who promoted the notion of 'unity in diversity, and difference in unity'. The meaning of any construct is generated by the complementarity of opposite poles. This corollary reminds the therapist to seek for implied or hidden contrasts about which the patient cannot be explicit. It is only by coming to understand what else the person might have been that we can make any sense of what he has in fact become. In developing his argument for 'descriptive complementarity', Varela (1979) comments:

> "*There is, evidenty, a need to overemphasize a neglected side of a polarity. Similarly, autonomy cannot in fact be conceived without a complementary consideration of how the system is also controlled in a dual context; in particular, autopoiesis and allopoiesis are complementary rather than exclusive characterizations for a system.*" (Varela, 1979, p. 71).

Relations: Those corollaries which make statements about the relations obtaining among the components now follow.

Organization Corollary: "*Each person characteristically evolves, for his convenience in anticipating events, a construction system embracing ordinal relationships between constructs.*" This spells out the fact that the system must be comprised of clear channels of inference and movement, allowing the resolution of crucial contradictions, paradoxes and conflicts. The system must have a network of relations among constructs so that we can move among the network in an organized manner. This is a network of implicative relationships liberating us so that we can flow along lines of inference from one construction to another. While these lines of implicative relationships generate the possibility of movement, at the same time they form a web of constraints outside of which we cannot move. Thus, the personal construct system is a hierarchically organized system, systematically patterned to minimize incompatibilities and inconsistencies in the ways we approach events. Within the hierarchical organization, the higher-order superordinate constructs tend to be increasingly abstract, value-laden, and invariant. The most superordinate are called core-constructs and these govern our identity and continuing existence.

Fragmentation Corollary: "*A person may successively employ a variety of construction subsystems which are inferentially incompatible with each other.*" This tells us that subsystems of constructs within the overall system do not have to relate to one another in a logically coherent manner. When we closely examine our construct systems we find hypotheses or predictions which are not derivable from one another. Degrees of inconsistency, self-contradiction and inferential incompatibility are allowed, or even more so, valued.

> "*For man logic and inference can be as much an obstacle to his ontological ventures as a guide to them. Often it is the uninferred fragment of a man's construction system that makes him great, whereas if he were an integrated whole - taking into account all that the whole would have to embrace - the poor fellow would be no better than his "natural self".*" (Kelly, 1970, p. 20).

There are three other corollaries commenting on construct relations i.e. Experience, Modulation and Choice, but I will deal with them under later headings.

Change and Stability

General Processes of Change: We have already noted that the Fundamental Postulate contains a specification of the person as a form of movement in continuous cycles of anticipation. Kelly construes processes as more basic than inert substances as the following corollaries outline.

The Construction Corollary helps to define how we engage in anticipating: "*A person anticipates events by construing their replications.*" Life only makes sense when we plot it along the time dimension. Earlier we saw the agreement between Maturana and Kelly regarding the anticipating of recurrences of events. In the ever changing stream of life's events we bring forth invented repetitions through our predictive processes. From this viewpoint the Construction Corollary helps to create stability in the midst of flux.

The Experience Corollary helps to elaborate the effects of our invention of recurrences on our construct system. It states that "*A person's construction system varies as he successively construes the replications of events.*" This underlines the evolutionary nature of the system. When our anticipation of events is invalidated or shown to be ineffective we are immediately invited to reconstrue. As we are exposed to invalidational perturbations in the medium our construct system will vary to compensate for these aspects of our experience. Maturana echoes this position:

> "*For a change to occur in the domain of interactions of a unit(y) of interactions without its losing its identity with respect to the observer it must suffer an internal change. Conversely, if an internal change occurs in a unit(y) of interactions, without losing its identity, its domain of interactions changes. A living system suffers an internal change without loss of identity if the predictions brought forth by the internal change are predictions which do not interfere with its fundamental circular organization. A system changes only if its domain of interactions changes.*" (Maturana, 1980, pp. 11-12).

If the person does not alter his construct system in the light of invalidation he cannot have 'new' experience but only the 'same' experience endlessly repeated. Unless internal change occurs he cannot put himself into a different embodied relationship with events. He cannot change his interactions.

> "*...the succession we call experience is based on the constructions we place on what goes on. If those constructions are never altered, all that happens during a man's years is a sequence of parallel events having no psychological impact on his life.*" (Kelly, 1970, p. 18).

Kelly explains that accuracy of prediction cannot be taken as evidence that 'one has pinned down a fragment of truth', but suggests that a better criterion of effectiveness is the imaginative opening up of novel dimensions of interaction.

> "*And yet, however useful prediction may be in testing the transient utility of one's construction system, the superior test of what he has devised is its capacity to implement imaginative action. It is by his actions that man learns what his capabilities are, and what he achieves is the most tangible psychological measure of his behaviour.*" (Kelly, 1969, p. 33).

Kelly's comment here about the "transient utility" of one's system of knowing is in entire concordance with von Glasersfeld's radical constructivist position of the viability of the system as something which gets by the constraints of the 'real'.

Stabilizing Change

Particular corollaries specify the limits to change within the system's way of evolving itself.

The Modulation Corollary introduces the notion of construct "permeability" as the central mechanism for controlling change in the system. This corollary proposes that "*the variation in a person's construction system is limited by the permeability of the constructs within whose ranges of convenience the variants lie.*" This places constraints on the degree of evolution suggested in the Experience

Corollary. Changes occur within the overall comprehensive framework of the system, and to the degree that the relevant superordinate constructs are permeable then the more likely it is that the system can accommodate changes of structure. In discussing the experience cycle we saw that the system must have the capacity to admit and integrate the revised construct which emerges at the end of the cycle. Thus, what Kelly means by permeability is not the 'plasticity' of a construct as such but "*its capacity to be used as a referent for novel events and to accept new subordinate constructions within its range of convenience.*" (Kelly, 1970, p. 19). Unless we can integrate the novelty into the system it is likely to remain ignored. Less formally Kelly defined a permeable construct as one which "*takes life in its stride.*" It allows new experiences to be added to the going system. By contrast, an impermeable construct is one which rejects new events purely on the basis of their newness. Examples of impermeability may be found among our more compulsive research colleagues who need to open a new file for each new variable (or experience) they encounter. In summary, permeability refers to the recognition of novelty ('outside') and to the integration of novel constructs within the superordinate hierarchy ('inside'). As a structural limitation this corollary implies that you can learn only what your framework is designed to allow you to bring forth in events. This is the fundamental constraint on how much structural change we can undergo before threatening our organization with disintegration.

The Choice Corollary unpacks further implications for the evolution of the construct system and constraints on the directionality of construing processes. It states that "*A person chooses for himself that alternative in a dichotomized construct through which he anticipates the greater possibility for the elaboration of his system.*" Kelly believes that the person will always make choices which will increase the usefulness of the system. This increase is achieved in two ways, by defining the system and by extending it. The definition of the system involves either clarifying how constructs are applied to events or by specifying how the constructs are related to one another within the system (network of implications). This is an attempt to consolidate one's system. Attempts to extend the system involves amplifying it to cover new areas of application. In this case the intention is to extend the range. Notice that 'making a choice' for Kelly refers to self-involvement and is not defined in terms of the 'external' object in question.

> "*So when a man makes a choice what he does is align himself in terms of his constructs. He does not necessarily succeed, poor fellow, in doing anything to the objects he seeks to approach or avoid. Trying to define human behavior in terms of the externalities sought or affected, rather than the seeking process, gets the psychologist pretty far off the track... So what we must say is that a person, in deciding whether to believe or do something, uses his construct system to proportion his field, and then moves himself strategically and tactically within its presumed domain.*" (Kelly, 1970, p. 16)

This is very similar to Maturana's notion of how the organization specifies the cognitive domain wherein we can act with reference to the conservation of our organization. The construct prescribes what the person does and not what the object does. Furthermore, the choice is between the alternatives expressed in the dichotomous construct and not between the events discriminated by using the construct. This re-emphasizes the closure of the psychological system in that the person engages in an internal conversation with themselves. The choice is also constrained by the dichotomy corollary's statement that the system is composed of a finite number of bi-polar constructs. This is similar to Maturana's notion of structure-determinism, since the person must choose from within the current structures composing his system. Further, in order to positively elaborate the system he must 'choose' those dichotomized alternatives which will lead to extension and/or definition of the system. Since these implicative pathways are already laid down, one could argue that the 'choices' are illusory since the structures of the system already contain the preferential direction of movement. Maturana is not a constructive alternativist because at the moment of choosing there are no other alternatives possible. The choice made was the choice determined by the system's coherence. It had to be made. Curiously, Kelly would appear also to be not an alternativist from this reading into his corollaries of choice, dichotomy and modulation.

There are many clinical examples of clients who present for psychotherapy precisely because their bipolar constructs (and their system of implications) offer 'choices' which are all negative. For example, the anorexic who fears to become obese, or the manic person who fears moving to depression again. Whichever pole of the construct they attempt movement through leads them to painful experience and a sense of being trapped. Thus, where you place yourself along the construct axis is not nearly so important as the fact that you have evolved that particular construct in the first place. Once it is part of your structure you are constrained by it. Thus, construct structure can be alternatively liberating or imprisoning.

The Cognitive Domain

Maturana uses the notion of the cognitive domain to characterize effective action of the system over time.

> "*A cognitive system is a system whose organization defines a domain of interactions in which it can act with relevance to the maintenance of itself, and the process of cognition is the actual (inductive) acting or behaving in this domain.*" (Maturana, 1980, p. 13).

From Kelly's point of view this involves the Modulation Corollary since the superordinate constructs and their relations relate to the organization of the system. It is the organization which specifies the cognitive domain, where our actions can be effective in the maintenance of autopoiesis. We have seen that a system undergoes change only if its domain of interactions changes. As we have also seen there are only two types of changes - changes of state and changes that destroy organization. The Modulation Corollary attempts to place limits on novel perturbations in order to conserve organization. This can be so successful that it leads to a rigidity of the system or what Kelly called the "hardening of the categories".

Dorothy Parker's version was that "you can't teach an old dogma new tricks".

The Range Corollary also helps to define the cognitive domain in terms of those events we can effectively interact with. It states that "*A construct is convenient for the anticipation of a finite range of events only.*" The issue here involves a question as to the number of events a construct can effectively deal with. The Range indicates those events which the construct allows one to anticipate and understand. Few constructs (if any) are relevant to everything, and certain features of Kelly's Theory lets us know when the construct system as a whole is out of its depth - these involve the transitional constructs which help us construe personal transitions or becomingness as our system moves into self-reconstruction.

There are three final corollaries relevant to the domain of interactions.

The Individuality Corollary underlines the idiosyncratic and autonomous nature of construing. It states that "*Person's differ from each other in their constructions of events.*" Kelly's view is that construing is so personal a matter that no two people are ever likely to invent identical systems. Furthermore, he suggests that 'even particular constructions are never identical events'. This emphasizes his view of continuing change within the individual's meaning system. This goes well with Maturana's statements about each observer bringing forth a different reality and not merely alternative 'versions' of the same objective reality. It also fits with the 'closure' thesis of Varela (1979) which leads to the abandonment of notions of 'instructive interactions' and 'programming' people. Varela notes that:

> "*...closure and the system's identity are interlocked, in such a way that it is a necessary consequence for an organizationally closed system to subordinate all changes to the maintenance of its identity.*" (Varela, 1979, p. 58).

To counterbalance this corollary, Kelly proposed the Commonality and Sociality Corollaries which brings us more into the domain of interpersonal role enactments.

The Commonality Corollary states that "*To the extent that one person employs a construction of experience which is similar to that employed by another, his processes are psychologically similar to those of the other person.*" This reaffirms the constructivist position that our actions are governed by our constructs. Insofar as people construct events in a similar manner they may behave in a similar manner to one another (irrespective of whether or not the events themselves are identical). It is what we make of events that matters.

> "*...the extent of psychological similarity between the processes of two persons depends upon the similarity of their constructions of their personal experiences, as well as upon the similarity in their conclusions about external events.*" (Kelly, 1970, p. 21)

Maturana's reference to this phenomenon is in terms of the structural intersection of systems. He says that

> "*when two composite unities structurally intersect through their components, they share components and have as composite unities the same domain of existence.*" (Maturana, 1987, p. 347).

Kelly's corollary of commonality does not say that the two people have to have experienced the 'same' events, nor even that their cycles of experience were similar. What must be similar is their construction of experience.

The final corollary is Sociality: "*To the extent that one person construes the construction processes of another, he may play a role in a social process involving the other person.*" Kelly here distinguishes between construing merely the behavior of the other person as opposed to the necessity of construing his construction processes. The construing of mere behavior (as if people were automata) does not generate social processes but rather lead to the development of manipulative devices which attempt to assert the myth of instructive interactions. The alternative is construing the other's construing processes where we *"strive for some notion of the construction which might be giving your behavior its form."* (Kelly, 1970, p. 24). The latter approach allows the emergence of a social role as opposed to merely 'doing' things to others.

> *"If I fail to invest in a role, and relate myself to you only mechanistically, ... I shall probably take my predictive failures as an indication only that I should look to see if there isn't 'a screw loose somewhere' in you."* (Kelly, 1970, p. 25).

This leads us into the final section of the chapter where we construe some common psychotherapeutic issues in Kellian terms.

Implications for Psychotherapy and Ethics

Within the constructivist model and particularly Maturana's notion of (objectivity) many conventional psychotherapeutic distinctions must either be redefined or dropped from our lexicon entirely.

Apostolic Zeal

One of the first issues that psychotherapists must deal with is what Michael Balint (1977) called 'apostolic zeal' and what Kelly regarded as the therapist's *"overwhelming desire to compel the client to get well"* (1955, p. 1194). Kelly's model of the psychotherapeutic relationship as a creative, co-operative, experiential and experimental process clearly implies that therapy is not something the therapist 'does to' the client. There are no mechanical technologies of influence 'applied to' the person. There is no linear causality that can dictate changes in another's system. Mistakenly believing that there is such causality often leads to therapist hostility toward the client. The broad aim of psychotherapy is to aid the client in precipitating a healthful psychological process at a more rapid rate of change than he might achieve from his own efforts. This is not to say that we aim to create a fixed state of mind or well-being by the end of therapy, but rather that the client has resumed his own growth process, i.e. his ability to conserve his organization and his adaptation (structural coupling) has been enhanced. The therapist's role is not to provide a blueprint of what the patient must eventually become but rather to guide the person in experimenting with some of his burning issues through a full cycling of the experiential cycle. The idea is not merely to get people back on their feet again, but rather to get them moving.

> "*...the purpose of therapy is not to produce a state of mind but to produce a mobility of mind that will permit one to pursue a course through the future.*" (Kelly, 1955, p. 208).

Since pathology is brought forth by the distinctions of the observer, so equally would the specification of 'health'. Having abandoned the myth of instructive interactions the constructivist therapist can also abandon the notion of being responsible for 'causing' the person to 'improve' or be 'cured'. Rather, therapy must be based on co-evolving a coherence with the client (getting into a co-ontogenic structural drift) in such a way that we provide an experimental context within which he may begin to change his way of changing to conserve his stability. Kelly agrees with Maturana that we cannot predict or predetermine the outcome of drifting in therapy.

> "*But the psychotherapist does not know the final answer either - so they face the problem together. Under the circumstances there is nothing for them to do except for both to inquire and both to risk occasional mistakes.*" (Kelly, 1969, p. 229).

Transference

Given Maturana's view of organizational closure and (objectivity) every description made by an observer is a 'transference'. That is we bring forth our own reality by projecting ourselves (our interpretations). For Kelly, the provision of a fresh set of elements is important in forming a novel context which may trigger the person to create new constructs. The therapist will frequently use himself as a fresh element to perturb the transferences of the client.

The therapist continually interacts in such a way as to extricate himself from the client's attempts to transfer onto him parts of his outmoded construct system. The client is thus 'invited' (triggered) to bring forth the therapist in alternative and novel ways. If the client were to succeed in dressing up the therapist as a figure from his past or present life, the therapist could find that he had unwittingly donned a strait-jacket and was now severely constricted in his capacity for therapeutic effectiveness. That is, he now helps to constitute the patient's organization by interacting with him through the crucial axes constituting the patient's construct system. The therapist becomes part of the problem and not part of the solution. For Kelly the issue of transference therefore involves the experimentation with role constructs. It becomes a problem only when the person continues to impose the same transferred construction onto the therapist and does not make something out of the disconfirming interactions of the therapist. It is a problem therefore of an inability to complete the cycle of experience and invent new constructions (cf. Kelly, 1969, p. 223).

Resistance

Many therapists who continue to believe in instructive interactions will inevitably encounter "resistance" or "stubbornness" in their patients who will be described as "refusing to see the point", or as "sabotaging directives or interventions", or as "deliberately not wanting to recover". However, from the constructivist point of view the person is doing the only thing that makes sense for their system to do. Their interactions are governed by their construct structures (structure-determinism) and as Kelly notes, he found that invoking the concept of 'resistance'

> *"bespoke more of the therapist's perplexity than of the client's rebellion....it was possible to see resistance in terms of the therapist's naivete."* (Kelly, 1969, p. 83).

Insight

In personal construct psychotherapy what is emphasized is getting the client into novel interactions (experimentation) and the purpose of therapeutic conversations is not to 'produce insights' but rather to produce movement along construct channels which will commit the client to constructive action. As von Foerster's (1984) aesthetical imperative states, "*If you desire to see, learn how to act.*" Therapeutic conversations are also forms of experimental (inter) action where experiences never before spoken about (brought forth) are encountered within the languaging of the consultation. Kelly recommended using a language which would never confirm or make anything certain (1969, p. 159), but rather would allow us to approach events propositionally. Therapists often become 'stuck' at a point where the client develops what is, from the therapist's point of view, a clear and accurate perception of hitherto confusing experiences. Several sessions later as the client continues to move through other levels of construction which supercede the previous "clear perception", the therapist is often found attempting to bring the client back to the 'truth of the matter'. Where the client shows (appropriate) irritation at being held back in his progress, Kelly notes that

> "*One of the most amusing yet baffling experiences in psychotherapy is the way today's "insights" can become tomorrow's "resistance". Psychotherapists often stand on their heads to retain what they once hailed as a remarkable insight in their patient's step-by-step analysis ... if he had regarded the client's new construction as a hypothesis rather than an insight in the first place, he could have saved himself a lot of anxiety once it became clear to both of them that the therapy must move on to other levels of construction.*" (Kelly, 1969, p. 159)

Within personal construct psychology, "insight" is another notion that has no place since the philosophy of alternativism says that any one event can be viably seen through many different "insights" each with their own validity. Thus, in the same way that all the observer's operations of distinction are transferences, so they are also all "insights". That is to say the experience of 'insight' is simply that something novel has been brought forth, i.e. that something new is 'in sight'.

Praxis

The important features in developing a conversational-constructivist model for psychotherapy are summarized as follows.

1) The notion of embarking in a co-ontogenic structural drift with the client.
2) The generation of a boundary to the dyad (or family) through the network of interactions between the participants.
3) The maintenance of this boundary.
4) The triggering of the observer role in the client system.
5) Leading to what Maturana calls 'reflection in the domain of action'.
6) The identification/bringing forth of the organization of the conversations which the client must constitute and by so doing becomes defined as a patient. ('Listening to the listening').
7) The identification of orthogonal axes of approach to the client's system.
8) Triggering changes of structure destructive of the current organization.
9) Promoting changes in the way the person changes himself in order to remain stable - i.e. dissolving (Vs solving) the problem.

A family organization can be identified in terms of the network of conversations which contains the relations of constitution of the family. Each family member is a component of the system who contributes the specific axes or construct dimensions which constitute the family organization. The only way to disintegrate this organization is through interactions which do not pertain to relations of constitution of the system, but rather encounter the components (individuals) in an orthogonal manner (i.e. through axes irrelevant to the constitution of the organizational invariance of the system). If we do not interact orthogonally we simply become an extension of the family problem and thereby help to constitute and validate its organization. The therapist must therefore offer a subtle blend of the intriguingly alien with the reliably familiar in order to be in a position to provoke the destruction of the self-maintenance constructs of the family and individual systems. To do this is to disconfirm the going system and to cause it to disintegrate. Initially this is done by composing an alternative lexicon for conversations taking place within the therapeutic system.

Kelly: Types of Change

Kelly points out that there are two main ways to change the construct system, namely to 'reroute' the person within existing channels or to invent new channels of movement. These two correspond to Maturana's two forms of structural change, i.e. changes of state and destructive changes. Kelly outlines eight (non-exhaustive) approaches to change within the psychotherapeutic context. Two of these seem most appropriate for having the potential to trigger a disintegration of organization and these two involve on the one hand, the alteration or redefinition of the meaning of a construct and, on the other hand, the invention of new dimensions for movement. Kelly did not much believe in attempting to 'repair' a broken down system but rather felt we should strive to create something new.

> "*As long as any client is inclined toward undoing the mistakes of his past rather than creating a constructive system which does not call for the repetition of those mistakes in the future, very little psychotherapeutic movement is likely to take place.*" (Kelly, 1955, p. 380).

The therapist's objective, therefore, is to promote the emergence of novel structures in the construing system which brings forth an alternative reality which does not support the presence of the previous problematic reality. That is, the person no longer makes the operations of distinction to bring forth the dilemma in a domain of existence within which he was previously trapped. It is no longer a focus of his anticipatory coherence. Indeed some therapists who remain unconvinced of their client's progress are often found 'prescribing a relapse' where the previous constructions are again 'exercised', presumably to exorcise the fears of the therapist as to the 'genuineness' of the client-system transformation.

Within the range of approaches invented by Kelly his method known as 'Fixed Role Therapy' most clearly illustrates his emphasis on orthogonality, reflection in the domain of action, and the increasing of the cognitive domain and domains of existence through the invention of a different identity. This approach involves getting the client to write a self-characterization which the therapist then analyses with the objective of producing a different invented character whose domain of existence is orthogonal to that of the original sketch. Recalling that we are a "multiverse" of selves, only certain structural features (components and relations) of each individual are required to be constitutive of the system and hence, there are many axes/dimensions which are 'superfluous', i.e. dimensions not constitutive of the system. It is with these structural dimensions that the therapist must interact, thereby remaining orthogonal. Whatever he does with the system structure must not confirm the organization.

In writing the invented character the therapist uses these orthogonal dimensions and may introduce novel dimensions also. If we imagine the current organization/structure of the client's system as a ship in a path of drift, then what we aim for is to get him to "jump ship" and take off in an alternative ship composed of a different (orthogonal) organization/structure and so begin to move in a different drift. Through enacting the invented role the client comes to specify and interact in a novel cognitive domain. Through changing his domain of interactions the person changes his system.

Maturana comments as follows:

> *"...there are as many domains of cognition as there are domains of existence specified by the different identities that living systems conserve through the realization of their autopoiesis. These different cognitive domains intersect in the structural realization of a living system as this realizes the different identities that define them as different dimensions of simultaneous or*

> *successive structural couplings, orthogonal to the fundamental one in which the living system realizes its autopoiesis. As a result, these different cognitive domains may appear or disappear simultaneously or independently according to whether the different structurally intersecting unities that specify them integrate or disintegrate independently or simultaneously...*
>
> *It follows from all this that a living system may operate in as many different cognitive domains as different identities the different dimensions of its structural coupling allow it to realize. It also follows from all this that the different identities that a living system may realize are necessarily fluid, and change as the dimensions of its structural coupling change with its structural drift in the happening of its living. To have an identity, to operate in a domain of cognition, is to operate in a domain of structural coupling*". (Maturana, 1987, pp. 361-362).

Concluding Comments

Individual responsibility is paramount in the constructivist model since we become aware that we are self-inventing it is our responsibility to be careful as to how we go about this task. Since nothing exists without the observer then also we are fully responsible for what we bring forth in our lives. Events have no separate existence apart from our distinguishing them. Thus when a therapist brings forth a 'family' it is his invention, and he must keep in mind that for each member of the family there is a different reality (family) brought forth. The temptation for the therapist to believe that his "family" is the 'really real' family must be abandoned along with any notion that he has a privileged access to independently existing reality. Maturana concludes that

> "*...all things, are cognitive entities, explanations of the praxis or happening of living of the observer, and as such, as this very explanation, they only exist as a bubble of human actions floating on nothing. Every thing is cognitive, and the bubble of human cognition changes in the continuous happening of the human recursive involvement in co-ontogenic and co-phylogenic*

drifts with the domains of existence that he or she brings forth in the praxis of living. Everything is human responsibility." (Maturana, 1987, p. 377).

References

Balint, M. (1977). *The doctor, his patient and the illness*. Kent: Pitman.

Eigen, M. (1971). Self-Organization of matter and the evolution of biological macromolecules. *Naturwissenschaften*, **58**, 465-523.

Foerster, H. von (1984). On Constructing a Reality. In: P. Watzlawick (ed.), *The invented reality*. New York: Norton.

Glasersfeld, E. von (1984). An Introduction to Radical Constructivism. In: P. Watzlawick (ed.), *The invented reality*. New York: Norton.

Glasersfeld, E. von (1985). Reconstructing the concept of knowledge. *Archives de Psychologie*, **53**, 91-101.

Glasersfeld, E. von (1987). *The construction of knowledge*. Seaside, Cal.: Intersystems.

Jantsch, E. (1980). *The Self-organizing universe*. Oxford: Pergamon.

Jantsch, E. (1981). Autopoiesis: A Central Aspect of Dissipative Self-Organization. In: M. Zeleny (ed.), *Autopoiesis: A Theory of Living Organization*. New York: North Holland.

Keeney, B.P. (1983). *Aesthetics of change*. London: Guilford.

Kelly, G.A. (1955). *The psychology of personal constructs*. 2 Vols. New York: Norton.

Kelly, G.A. (1969). Man's construction of his alternatives. In: B. Maher (ed.), *Clinical psychology and personality; The Selected Papers of George Kelly*. New York: Wiley.

Kelly, G.A. (1970). A Brief Introduction to Personal Construct Theory. In: D. Bannister (ed.), *Perspectives in personal construct theory*. London: Academic Press.

Kelly, G.A. (1977). The Psychology of the Unknown. In: D. Bannister (ed.), *New perspectives in personal construct theory*. London: Academic Press.

Maturana, H.R. (1978). Biology of language; Epistemology of reality. In: G.A. Miller, E. Lenneberg (eds.), *Psychology and biology of language and thought*. New York: Academic Press.

Maturana, H.R. (1980). Biology of cognition. In: H.R. Maturana, F.J. Varela, *Autopoiesis and Cognition*. Boston: Reidel.

Maturana, H.R. (1987). The Biological Foundations of Self-Consciousness and the Physical Domain of Existence. In: E.R. Caianiello (ed.). *Physics of cognitive processes*. Singapore, New Jersey, Hong Kong: World Scientific.

Maturana, H.R. (1988). Reality: The search for objectivity or the quest for a compelling argument. *Irish Journal of Psychology, Special Issue on "Radical Constructivism, Autopoiesis and psychotherapy"*, V. Kenny (ed.) **9**, 1, 25-82.

Mendez, C.L., F. Coddou, H.R. Maturana (1988). The bringing forth of pathology. *Irish Journal of Psychology, Special Issue on "Radical Constructivism, Autopoiesis and Psychotherapy"*, V. Kenny (ed.), **9**, 1, 144-172.

Polanyi, M. (1958). *Personal knowledge*. London: Routledge & Kegan Paul.

Prigogine, I., I. Stengers (1985). *Order out of chaos*. London: Flamingo.

Varela, F.J. (1979). *Principles of biological autonomy*. Oxford: North Holland.

6

The Art of Self-Management

Kim James

Abstract. This chapter deals with the use of the graphic symbolizing process in psychotherapy. It starts from the theory of Direct Visual Perception as a synergistic process where perception is the value-state of relations holding between the organism and its environment, Hegel's "spiritual animal kingdom". The graphic act differentiates the aspects of this perception in symbolic value-relations which modify the artist and enable the differentiation of the individual in the group as Self and Other. "'I' that is 'We' and 'We' that is 'I'". The first part of this chapter presents some of the basic concepts from the work of J.J. Gibson on perceptual and cognitive faculties. The second part relates these notions to the practice of psychotherapy, in particular to a psychotherapeutic technique which makes extensive use of painting and other graphic activities.

IN THIS CHAPTER we discuss the Self as the result of a self-organizing system. What sort of system is it which organizes a Self? A self-realizing system. The Collins dictionary defines self-realization as: The realization or fulfillment of one's own potential or abilities. Which, it seems to us, is what psychotherapy is concerned with.

It is necessary at the outset to state that it may be that we take a radically different stand to the majority of the participants in this symposium on the nature of the relations between the Self and Reality. The most extreme view of the relations between the Self and Reality, states that there is no objectively existing reality independent of any observer; Whilst others think that the most we ever achieve is to discover what the world is not.

We take the point of view that the Self *is* a manifestation of reality, there is no point in taking a dualist stand that there is some gap between a hypothetical "inner world" and a hypothetical "outer reality".

> "*The supposedly separate realms of the subjective and the objective are actually only poles of attention. The dualism of observer and environment is unnecessary. The information for the perception of "here" is the same kind as the information for the perception of "there"*." (Gibson 1979.)

The problems associated with the Self and their effective resolution are not open to adequate description through talk of 'inner processes', whether this is in the terms of psychoanalysis or variations of structuralism. We have chosen to follow the lead put forward by J.J. Gibson (1966) in looking at the senses considered as perceptual systems, since it gives us the possibility of intervening in the therapeutic process in a genuinely objective fashion. Our starting point is the theory of Ecological Psychology put forward by J.J. Gibson (1979). As such it is a therapy which has as its basis a theory of Direct Perception. The theory of Direct Perception as put forward in Gibson has the great merit of stating clearly the material basis of the Self.

> "*The experience of a central **Self** in the head and a peripheral **Self** in the body is not (therefore) a mysterious intuition or a philosophical abstraction but has a basis in optical information.*"

> "*I have described this information for perceiving the **Self** in terms applicable to a human observer, but the description could be applied to an animal without too much change,...... Each species sees a different **Self** from every other. **Each individual***

> *sees a different Self. Each person gets information about his or her own body that differs from that obtained by any other person.*" (Gibson, 1979, p. 115)

We follow Gibson (1979) in acknowledging the dominant role of the perceptual systems in the Knowing system, but we consider the latter also as a system in its own right, in our approach we attempt to follow the same descriptive reasoning as Gibson(1966).

We do not accept an internal construction of meaning on the part of the person, for us meaning is the perception of the affordances of material existence found in the synergistic function of the entity in environment. Existence is meaningfulness and is not inside or outside of the perceiving entity. What is specified in languaging is those *aspects* of the totality of the existence which we *distinguish* and *name* as *Meaning.*

> "*There has been endless debate among philosophers and psychologists as to whether values are physical or phenomenal, in the world of matter or only in the world of mind. For affordances as distinguished from values the debate does not apply.* ***Affordances are neither in the one world or the other inasmuch as the theory of two worlds is rejected.***" (Gibson, 1979, p. 138)

Gibson created the word 'affordance' to signify the possibilities for action which the world affords to a living organism. An affordance is perceived directly and is species and development specific. This process of perceiving an affordance is not one of perceiving a value-free physical object to which, Gibson points out, meaning is somehow added in a way which nobody has ever been able to explain; but it is a process of perceiving a value-rich ecological object.

> "*...if there is information in light for the perception of surfaces, is there information for the perception of what they afford?....If so to perceive them is to perceive what they afford. This is a radical hypothesis, for it implies that the "values" and "meanings" of things in the environment can be directly*

> *perceived. Moreover it would explain the sense in which values and meanings are external to the observer.*" (Gibson, 1979, p. 127).

Gibson points out that in a system approach the arrows indicating action on a system point both ways, to the environment and to the observer. All perception of '*where* one is' is accompanied by a perception of 'where *one* is'. Self-perception and environment perception are two sides of the same perception, they are only available to inspection as separate entities through the symbolism of the thought processes of the Knowing system.

> "*An affordance, as I said, points two ways, to the environment and to the observer. But this does not in the least imply separate realms of mind and matter, a psychophysical dualism. It says only that the information to specify the utilities of the environment is accompanied by information to specify the observer himself, his body, legs, hands and mouth. This is only to re-emphasize that exteroception is accompanied by proprioception that to perceive the world is to co-perceive oneself. This is wholly inconsistent with dualism in any form, either mind-matter dualism or mind-body dualism. The awareness of the world and of one's complementary relations to the world are not separable.*" (Gibson, 1979, p. 141)

> "*In my view, proprioception can be understood as* ***egoreception,*** *as sensitivity to the* ***Self,*** *not as one special channel of sensations or as several of them. I maintain that* ***all the perceptual systems are propriosensitive as well as exterosensitive,*** *for they all provide information in their various ways about the observer's activities....*". (Gibson, 1979)

The Meaning which is directly perceived is constituted of values. Since perception is a synergistic process the perception of a value-rich environment is at the same time a perception of a concomitant self-value within that environment. The human being, as a total system, cannot perceive anything which she cannot perceive in transaction with an environment. So the question arises as to how she perceives herself as separate from others and from herself when she observes herself. The problem of the human condition is that of self perception as a value-rich entity whose value may be simultaneously established and denied in the human society by which she or he is mutually defined as human. It is only humans who appear to have an ability to know themselves as an entity with an identity that is to say to consider identity as if it were separate, to know themselves as their own object. The Self knows itSelf.

Dynamic analysis does not permit the concept of any process as isolated. This means that there is no place for terms which have arisen to convey ideas of separate entities such as 'mind' and 'consciousness'. The common use of these terms conveys the impression of processes which are peculiar, of a different order or degree from others with which they are nevertheless somehow related.Living systems are enduring conjunctions of mutually defining relations which have what an observer calls a plane of behavior proper to the specific control state process. Underlying all, however, are the properties of matter in motion which form reality. All the complexities of human interaction have to be understood in the same dynamic terms as any investigation of material processes.

Humans have a habit of abstracting the world into 'things' and 'forces' and our failure to see that a 'thing' is as much an expression of value transactions as the force which is said to move them tends to obscure a vision of the continuity of matter.An examination of the fundamental particles shows that we cannot use expressions such as 'consists of' in any sense of Newtonian 'hard massy particles' (Bohm & Peat, 1987). Instead we have to consider units of power constrained into limited endurance by mutual defining influences of power. These units of power are any "thing" from the smallest particle, through people, to the Universe.

The power relations which underlie structure at all levels are nested formal relations which unfold, somewhat in the manner of Mandelbrot's fractals (Bohm & Peat, 1987). It is the observer who, at specific distances of her power relation to the entity, denotes the structure as having qualities in relation to her own powers to do.

If matter is taken to be units of power, (ability-to-do in diversities, limited in amount and different in direction) these diversities in their mutual influences define systems, associations with identity, which in relation to other associations, maintain spatial and other relations. The energy transfers which are necessary to maintain particulate identity take values according to what attractions and repulsions are necessary to ensure endurance. Living systems have forms where the endurance in stability is only maintained by increasingly complex conjunctions.

The Knowing system is the control condition for all the systems of the organism in their mutual transactions. It is defined into existence by the necessity of these nested systems to resolve their conflicts of interest in their exchanges with each other and the environment which is external to them all. It is the mutual constraint condition in transaction, which has a management function. The Knowing system is the system which is the condition of compromise for all the systems which interact. It does not have direct inputs from the environment itself but many of its contributing sub-systems do, for example the perceptual system and the digestive system. Similarly it does not have direct outputs to the environment but uses elements of the operating characteristics of several systems, digestive, respiratory and so on.

It is impossible to understand the Knowing system without placing it in its evolutionary gradient. There is a common tendency to consider the system as a fully formed entity and to ignore its evolution as a function. Each different system has evolved through the constraints of mutually defining relations within its niche to maintain its own circularity. This is as true for a system within an organism as it is for a species. The visual system does not have the "purpose" of informing the organism of events in the visual part of the electromagnetic spectrum, it transacts with photons, among other elements, to maintain its own circularity. In so doing it enables other systems to maintain theirs - it is constrained to *have* a function, as distinct from *functioning*.

> *"...the circular organization implies the prediction that an interaction that took place once will take place again.....In a continuously changing environment these predictions can only be successful if the environment does not change in that which is predicted. Accordingly, the predictions implied in the organization of the living system are not predicted of particular*

> *events, but of classes of interactions. Every interaction is a particular interaction, but every prediction is a prediction of a class of interactions that is defined by those features of its elements that will allow the living system to retain its circular organization after the interaction, and thus, to interact again. This makes living systems inferential systems, and their domain of interactions a cognitive domain.*" Maturana (1980)

As we understand Maturana he takes a monist standpoint that all information as to the "outside" world (if it exists at all) is the result of "internal" processing of otherwise meaningless inputs. From the logic of the statement which we quote above however it is possible to arrive at, what is for us, a much more satisfying conclusion.

> "*In a continuously changing environment these predictions can only be successful if the environment does not change in that which is predicted.*"

What this must mean is that there is a maintenance of order in the nature of the environment external to the organism some invariant features of the environment which under transformation (change) are responded to in terms of classes of relations. This can be nothing else but a structuring of phenomena which is independent of the organism. Reality exists therefore independent of the organism and this reality has its own structure. Can this reality be perceived or must we continually construct it?

Let us at this point state that we consider that reality can never be known in the sense of any complete Knowledge. Phenomena exist which we are quite unaware of at this time, as the progress of science and art continually show. Each time a discovery of new organizations of power is made we name it. The phenomenon of naming however is not the bringing into existence of reality but its distinguishing into aspects. We agree with Maturana that descriptions of the world are couched in terms of our own knowledge of our power to effect change on what we by distinction name as the outside world. However we feel that it is possible to draw a conclusion of the perceiving entity being itself a constituent of a reality which constitutes the meaningfulness of which the observer makes distinctions. There is a difference between the meaningfulness given in direct perception and the categories of meaning which constitute mediated perception and which are named.

Maturana points out that the cells which constitute the sensory surfaces are constituted of collections of cells with similar though not identical properties. Thus the interaction with the environment is through the classes of properties of the receptor cells, which in the vertebrates with their highly developed nervous systems makes the modification of the organism one of "pure relations" (Maturana, 1980).

> "*One thing is certain. The generally accepted picture of a sense organ as a mosaic of energy transducers each connected to a distinct nerve fibre and thereby to a distinct cell in the brain is quite wrong. In the skin, the retina and other organs, the receptive units constitute overlapping fields of cells....They are functional, not anatomical units. These units seem to modify their input as a function of a change of energy (sequential order) or a relation of energy (adjacent order),not as a function of the application of energy. What they register is not the energy of stimulation but the sequence or arrangement of stimulation, that is information.*" (Gibson, 1966)

This information is the classes of relations under transformation which Maturana describes. The receptor cell mosaic responds invariantly in terms of classes of relations to the invariant ordering of the relations of energy on the external side of the boundary. Thus there is no need for "processing" in the brain, what is necessary is an invariant system response, a requisite variety of states of systems capable of altering relationally as the relations at the interface change. What an animal perceives depends upon the evolutionary history of its species in environment. The evolution has been an endurance of a system of energy-values in transaction. All actions of the organism have had one direction toward balance, this is the necessary achievement if it was to survive.

The action of a multifarious environment tends towards the creation of new organismal states and determines a direction towards complexity. In this the anatomical structure called the nervous system is a mode of action directed to balance. It can only be properly conceived of as fulfilling a condition of balance in the mutual influences defining of energy-values transactions with the other systems constituting the organism. It does not choose to be a nervous system, it has no "purpose" (except to an observer as an abstraction) it has evolved toward balance by transaction with environment. Structure must be considered to be a manifestation of action and its building an achievement of action.

> "*Structure is often treated as being static and more or less complete in itself. But a much deeper question is that of how this structure originates and grows, how it is sustained, and how it finally dissolves. Structure is basically dynamic, and should perhaps better be called **structuring**, while relatively stable products of this process are **structures**.*" (Bohm & Peat, 1987)

The secondary status of form must be consistently acknowledged.

The energy-values which constitute the nervous system may be activated variously as one system or another and it is at all times a complex of spheres of relative activity. To this extent the units of function which constitute the perceptual systems are bound to anatomy and the relational transactions of energy-values. What has to be understood is:

> "*...for an animal to discriminate objects visually the receptors in its eyes must absorb light quanta and be activated; yet, the objects that the animal sees are determined not by the quantity of light absorbed, but by the relations holding between the receptor induced states of activity within the frame of the functional organization of the retina. Moreover since the domain of interactions is defined by its structure, relations with which the nervous system interacts are defined by this prediction and arise in the domain of interactions of the organism.*
>
> *The organization of the living system defines "a point of view", a bias or posture from which the interactions take place, and*

> *determines the possible relations accessible to the nervous system.*" (Maturana, 1980)
>
> "*The point of observation for an ambient optic array [must be] thought of as a position at which observation could be made, a position that could just as well be occupied by any observer, the invariants of the array under locomotion can be shared by all observers ...* ***the point of observation is public, not private.*** *When a point of observation is occupied, there is also optical information to specify the observer himself, and this information cannot be shared by other observers. For the body of the animal who is observing temporarily conceals some portion of the environment in a way that is unique to that animal, I call this information propriospecific as distinguished from exterospecific, meaning that it specifies the* ***Self*** *as distinguished from the environment.*" (Gibson, 1979, p. 111)

(For animals other than humans this state of the world corresponds to what Hegel calls the "spiritual animal kingdom" a society composed of separate individuals, each looking after his own affairs."This individuality which takes itself to be real in and for itself". (Hegel, 1977)

We have already pointed out that Maturana's formulation of prediction implies an existence of order in the universe external to the organism, a point which Gibson has already elaborated in his theory of Ecological Optics (Gibson 1979). The organism/ environment situation is synergistic. The domain of interactions is defined by the structure of the nervous system, but only in so far that this structure has a requisite variety of system states capable of reacting to the ambient flux of energy states in relation which arrive at the receptors. The structure of the nervous system has evolved in the mutual defining of the synergistic interaction of the organism in the environment and the domain of interactions of the organism is with the ordered classes of relations of the environment. The nature of evolution constrains the organization of the system and in so doing determines the possible relations accessible to the nervous system.

The synergistic transaction state of organism/environment with ordered invariant states on both sides of the boundary is perception. Humans can specify this state to themselves as if it were separate which is the operation of the Knowing system manifested in Thoughts.

The faculty of Knowing has been acknowledged to exist for a long time. The similarities of behavior of many animals leads to the inference that they share the faculty of 'Knowing' with man. It is frequently thought of as an information processing ability. In order to distinguish this as a system rather than as an internal processing mechanism we prefer to name it the Knowing system rather than use the current term cognitive or cognition. The cognitive state is that state which defines the process of living systems. The Knowing system is the internal state of recognizing the existence of self as entity in transaction and is a generated distinction. Living systems are enduring conjunctions of mutually defining relations which have what an observer calls a plane of behavior proper to the specific balance state process. Underlying all however are the properties of matter in motion which form reality. All the complexities of human interaction have to be understood in the same dynamic terms as any investigation of material processes.

The increasing complexity of the environmental field when members of the same species form a large (dominant) part increases the influence of values in transactions between the congregates of systems and the field. Where the field contains identical systems as part of the field the defining mutual influence constrains the internal systems of each contributant. The group which is constrained into endurance as a condition of its necessary transactions, acts as a single entity and part of its transactions may be named as a perceiving function. The constituent elements of the group system are subject to the constraints of the overall group perception of the group identity. The perception of the individual contributant is constrained by the group perception to see the Self as Other but Same.

In Hegel's terms: "'I' that is 'We' and 'We that is 'I'". Hegel defines Spirit as both the object which is known and the subject which gains knowledge of it: all knowledge for Hegel turns out to be self-knowledge. The criteria for knowledge cannot, for Hegel, be a standard outside the object but has to be contained within this object. The logical categories are categories of being and thus of becoming. They are not fixed, abstract forms, but concrete universals, knots in the web of history (Hegel, 1977).

Man's cognitive abilities represented in the Knowing system, are the survival of the necessary orderings of value systems in an eternally changing environment. They do not represent any achievement other than the requisite survival. There is little doubt that it was human social activity which played a major part in the evolution of the particular cognitive faculty which stamps him apart from the other animals, and in this activity stone tool making would appear to have played a major part.

The constraint on the interchange of energy values which define the animal kingdom which enables the individuation of man is found in the exchange of material objectifications of values as symbols. (cf. Baudrillard, 1972).

> "*Soon the matter has turned in such a way that as an individual he relates himself only to himself, while the means with which he posits himself as individual have become the makings of his generality and commonness*". (Marx, 1973)

Activity is the exchange of energy-values, which is a property of matter in its transactions. In the living system it is the condition of existence. Man in becoming a tool-maker objectifies the already existing digital quality of Knowing and by this objectification opens the way to the symbolism of Knowing as Thought. The general characteristic of living systems in interchange processes is that of analogue behavior. At any stage of an analogue process, however, it is open to description as a series of discrete digital steps. For the Knowing system to be able to describe itself to itself to think it must have a function which itself generates a digitalization of its process.

The Knowing system as the system which is the condition of compromise for all the systems which interact between themselves is constrained by its input constituents to be digital in its process. As the intensities of transaction of its constituents rise into primary influences on the system so it causes previous influence to diminish.

Given the wide possible inputs the transactions of interconnecting phase effects give discrete value entities as part of the condition of operating. The Knowing system therefore operates, in the higher animals and in man in particular as a linear sequential system; one effect must follow another, seen in the attention-mode of transacting, attention can only be given in order of priority. The fact that in common with all living systems it must be circular in its process means that it must obtain an ordering of the linear sequential operating towards closure. Thus it must achieve logic as an operation.

It is incorrect however to invoke tool making as a special kind of transaction, which of itself determines the process. Man became a tool-maker as a result of his interactions over an immensely long period and, we would venture to suggest, as a result of constraints on behavior which forced the adoption of new variants of previously existing behavior. In this development there was no miraculous discovery on the part of the being who was to become man, but a qualitative point at which the constraints in behavior held development into a structural stability of organization.

> "*The richest and most elaborate affordances of the environment are provided by other animals and, for us, other people.....They are so different from ordinary objects that infants learn almost immediately to distinguish them from plants and non-living things. When they are touched they touch back, when struck they strike back; in short they interact with the observer and with one another. Behaviour affords behaviour.*" (Gibson, 1979)

The features of 'attraction', 'repulsion', 'sharing' and so on, in transactions of powers-to-do, are the properties of material interchange which, in the form of human organization of symbolic object interchange, enable us to specify our interactions to ourselves to know, and to specify ourselves to ourselves as if we were Other to know that we know.The specification of Self as if it were Other is reflexive, in that Other is specified as if it were Self. Hence powers of relation can be ascribed to that which is not Self. Powers of relation which we perceive for ourselves can be ascribed to that which is not Self and vice versa through the mutual specification of self in environment as given by the affordances.

The perception of the newly emergent human species, whilst deriving from the commonality of act provides a perception which can be both a unique point of view and at the same time understood by the group as being both individual and group. The properties of the regulatory exchanges of the sub-system of the group are revealed to the observer (objectively) as the operation of selective transaction. Since perception is accompanied by an awareness of Self (Gibson, 1979), the perception of the object as meaningful gives meaning to the Self, objectively represented in the object, as meaningful to itself, as if it were Other.

At this stage the Self is conscious of itself as a member of the group but the individuation of the Self-as-individual, a member of the group but distinct, has yet to arise.

Self-identity is a two part construct, that given in perception is diversity in unity, that given in proprioception is of power-to-do.

> "*In my view, proprioception can be understood as* ***egoreception****, as sensitivity to the* ***Self****, ... the point I wish to make is that information about the* ***Self*** *is multiple and that all kinds are picked up concurrently.*" (Gibson, 1979)

Under normal conditions of function perception (which includes proprioception) is undivided and hence we have a sense of Self as whole. We have, however, learnt to divide our perception into aspects of a whole to our great advantage and also occasionally to our disadvantage. In cases of mental disturbance it is the sense of a divided or threatened Self which is at the root of the problem.

Those changes of state of the human action system which report themselves by the changes of state of the Knowing system are what humans term Ideas. These changes of state are frequently registered as emotion. In the absence of a symbolizing function the Idea is "formless", it is pure relation in Maturana's terminology and Meaning in the terms used by Gibson (1979). In our definition Thought is the output of the Idea in the symbol mode, and as such has a formal structure, since it is constrained to be linear, logical and sequential.

The distinction which we are making between Ideas and Thoughts is that Ideas are not form-bound, they are rarely linear, logical and sequential, and they do not have reference to the state of the individual Ego in the same structured way as the Thought. Ideas are synergistically referential and do not carry the reference to the Ego as split apart from its embedding. Thoughts on the contrary are referential to the Ego in relation to the environment. In short Ideas are *of* the world, Thoughts are *about* the world. It is this difference between the Idea and the thought which explains the type of confusion experienced in some types of migraine where the object is seen clearly but not understood in its context, similar also to the confusion described by Sachs in the case of the patient who mistook his wife for his hat.

It is important to realize that ideas are not formulated in words, or any externally referential symbol; they are difficult to describe, and attempts to do so are in word/symbol descriptions of internal "images" generated by one of the input systems of the Knowing system. The managerial nature of the Knowing system is demonstrated in the nature of ideas. Ideas arise as emotions from the disturbance of the interactions of the various systems constituting the bodyhood (this word which we came across in Maturana (1988) is an excellent description of the synergistic function of the organism in environment). They constitute a conversational domain for the systems and as such they cannot describe themselves to themselves. This constrains the process to attend to itself either through a sequence of actions or through the symbolic mode. Action on the environment is a metaphor of the emotional state of the bodyhood in the same way that the symbolic Thought is.

> *"The words or the language, as they are written or spoken do not seem to play any role in my mechanism of thought. ... combinatory play seems to be an essential feature in productive thought ... the elements are, in any case of visual and some of muscular type. Conventional words or other signs have to be sought for laboriously only in a secondary stage, when the mentioned associative play is sufficiently established*

> *and can be reproduced at will.*" (A. Einstein, in: Koestler, 1964)

A dream is a example of the functioning of the Knowing system, as is a 'feeling '. The dream state is one where the transactions of the systems report themselves in the mode of the visualizing system, primarily. Dreams are notoriously non-linear, non-sequential and non-logical. The same is true of the visual imagery commonly present in the waking state. It is the power of the transactions of the visualizing system which has given rise to the power of the word image in reporting the transactions of the knowing system. To confuse the functioning of the knowing system with linguistic ability is an error.

Ideas arise from two constituents, identifications of Self and presentations of Other which are given through the perceptual systems (Gibson, 1966). The identifications of Self may be presented in images as if they were Other, through the Knowing system. Presentations derive from transactions with the environment. The difficulties of ideas of Self when presented in images are well demonstrated in the case of Anorexia Nervosa where there is a deformation of the body-image. Identifications of Self have a derivation which is without an intervening medium, they are identifications of Unity, Power to do, Action, Effort and Endurance. Presentations of Other derive from transactions with the field. The Idea arises from the interactions of the various action sub-systems and is presented in the form of image which derives from the influences formed by the contributing power of the entities in the field, through exteroceptive and the proprioceptive sub-systems of the perceptual systems (Gibson 1966).

In humans the output is in word/symbols which convey the report of the system state in a manner which is communicable over distance and time in a linear sequential mode, enabling logical particular decisions to be made about the general state, both by the individual and the group system.

In language this is expressed as the Thought, "I must do (something)". All Ideas are predictive of activity. "I" is the expression of a power-to-do, as able. This primary Idea of self is present before speech. The Idea is one of influence but it is also conjoined with the Idea of self as material body.

> *"I have described this information for perceiving the **Self** in terms applicable to a human observer, but the description could be applied to an animal without too much change,.....They have radically different fields of view; their noses are different, and their legs are different, entering and leaving the field of view in different ways. Each species sees a different **Self** from every other. **Each individual** sees a different **Self**. Each person gets information about his or her own body that differs from that obtained by any other person."* (Gibson, 1979, p. 115)

Perception gives a knowledge of the unit of self as having "form", "substance", and "weight" in the same way as all reality which surrounds us. These elements are human ideas, presented in the form of Thought as word/symbol images of organizations of power-to-do with endurance. The scale of which is given by humans based on their own knowledge of their powers. "Thinking" is a word/ symbol for a special kind of power known to the individual who is acting in this mode. The ability to divide material existence into aspects labeled "past", "present" and "future" is a product of image symbolizing in the human. There is the Thought of "Time", coming, *from* the past *into* a present and on *into* a future. Properly speaking it is endurance which is perceived, it is the symbolizing faculty which enables divisions of endurance to be labeled in this way.

The power of the entity to act as a unity with endurance in the environment enables a self-perception as a dynamic system in transaction with a variety of continuing "doing" or processes of possibilities. This in the group animal brings the idea of dynamic relation. One of the first ideas which humanity has of itself is of dynamic relation.

If we combine all these self-referential ideas given by transactions action, endurance and dynamic relation, we can state that Knowing must be a state of dynamic process with relations, which derives from the transactions with energy states necessary to maintain the circularity of the organization of the entity in environment synergy. The Knowing system is a good example of the way a system is constrained into existence by the mutually defining influences of systems in transaction.

An analysis of the nature of the Ideas presented as Thought by the Knowing system shows that the system transacts with all the action systems of the organism. As such the ideas represent a management mechanism for the totality of the organism. Until we have a more extensive knowledge of the function of the system we have to limit ourselves to stating that the function is necessarily congruent with what the process receives. Many of the action systems it receives from are fairly narrow in their fields of exchange, the variety of the outputs of the Knowing system come from the fact that it is linked to a wide variety of states of organization and it is an extension common to many that are different in their powers of effect. The variety of output however does not preclude its naming as a specific process since the general output is uniquely represented in terms of accounting for power transacts.

The various action systems which form the living system do not transact with themselves. Which is why one cannot actively know one's knowing, any more than the digestive system will digest itself. In order to know that we know there must be an output of the Knowing system such that the output represents the state of the system and this output re-acts onto the system as if it were another Knowing system; that is to say that the representation to the system must cause another idea arising from the form of the output.

The activation of the Knowing system is provided by the inputs from any part. These inputs are a consequence of the inability of one or more given sub-systems to transact satisfactorily over a time to maintain their circularity. In these cases there is an imbalance which is conflictual with performance, such that other systems with which there is a transaction are themselves hindered. Signaling to the Knowing system brings about an imbalance of transaction within *its* power state. Man's cognitive abilities represented in the Knowing system, are the survival of the necessary orderings of value systems in an eternally changing environment. They do not represent any achievement other than the requisite survival.

We have already stated that what is perceived is Meaning on which humans confer the names of structures with functions when there is an endurance of invariant relations over time. A functioning of the Knowing system found in humans as a result of the existence of the symbolic function which mediates perception enables the separation of structure and function.

The Knowing system functions when contradictions arise from the transactions between the various action systems of the total entity which have to be resolved by changing the transactional state of this total system to favor one or other of the action sub-systems to maintain its circularity. *When everything is in non-contradiction there is no thought about it.* Since a system does not transact with itself, so the Knowing system cannot know itself. However in humans the symbolizing function enables the Knowing system to think about its thoughts as if they were Other. Without a self-referential symbolizing function this is impossible.

The Image/Symbol Process in the Therapeutic System

In humans there are several symbolizing functions, behavior itself can be used in a symbolic fashion, which has given rise to drama, music symbolizes the auditory/sound making processes and in the graphic process there has come into being one of the most profound areas of behavior due to the particular nature of linear trace making.

Drawing as a structured system is constrained into existence by the Mutual Influences Defining of systems with power in interaction. It is a structure which serves the balance condition of the interacting systems, it is not an invention of the 'human mind'; to a large degree it would be more correct to say that the activity has invented the human mind.

The progressive record of the movement of the hand leaving a graphic trace is qualitatively different from other symbol communication, the trace is permanent, lasting even beyond the lifetime of the maker. (We are talking here of the pictorial activity as distinct from its later development as writing). The trace communicates its maker's value state, in transaction of values, to the future. But the state it conveys is not explicit. It cannot be formulated in words. Most of the formless invariants in the array from even a complex pictorial relationship cannot be put into words. They can be captured by an artist but not described.

> "*....the essence of a picture is just that its information is not explicit. The invariants cannot be put into words or symbols. The depiction captures an awareness without describing it. The record has not been forced into predications and propositions. There is no way of describing the awareness of being in the environment at a certain place. Novelists attempt it, of course, but they cannot put you in the picture in anything like the way the painter can.*" (Gibson, 1979)

The perception of the trace would seem to be a powerful generator of pleasurable sensation, judging by the way the experience is repeated. In this way there is an enforced attention to the self in action and self-act as other (in the lasting trace). This is the start of the self-as-observer. It is a self-steering act. By this act the young child perceives an objective self-value, since in the reciprocal transaction of values which constitute perception the affordance points two ways and the meaningfulness of the action-Self is perceived long before there can be any words to express this quality.

Attention to the linear traces is accompanied by attention to the self as a value. From self-awareness there is a progression to attention to the Self as a worth for attachment. It is this Self-reference as worth for attachment, this objectified action-self which, we believe, is a more accurate description of that ill-defined concept which plays such a large part in psychodynamic theories, the Ego. Since there is a difference in perception between the perceived object as an invariant structure of meaningfulness and its representation in the graphic trace, attention is constrained to the difference/sameness ratio as meaningful in itself.

In the process of choosing the series of relational lines which will invariantly express the desired pictorial object the artist is obliged to pay attention to the object, in terms of variant aspects of angularity, occlusion and so on. The meaningfulness of the environment/artist perceptual synergy then includes attention to aspects of the artist, as much as the artist attends to aspects of the pictorial scene. "*The picture paints me as much as I paint the picture.*" (Picasso)

The awareness of self-in-action which accompanies all perception constraints the perceiver to an awareness of the meaningfulness of the Self-in-action as distinct and constituted of aspects. The artist is thus constrained into being a self-observer constituted of aspects and no longer an undivided whole. Self-perception and environment perception go together.

The mark made once has a value. It is a record of fulfilled intention on the part of the maker and thus specifies value for the maker as self-reference. The mark made again has added value, it is not just a pair of marks. "The value of the mark is the mark made again", to paraphrase Spencer Brown. This value is not the value for the maker in the way that her first one was. The value for the mark is in the relation between the marks, and this is the affair of what is now a drawing. A drawing is a set of marks which indicate value relations proper to the drawing itself. A drawing is an autonomous entity. All marks made on the drawing surface are made with reference to the marks previously made. The "intention" of the mark maker is modified by the trace she makes towards closure. A succession of marks made at random move rapidly towards closure. Closure is obtained when the structure of the drawing does not permit of any further modification. The drawing is then said by the mark maker to be "finished".

The power of the graphic act lies in the fact that the line is a manifestation for the human of an action carried out. It is a signifier of action. A combination of lines however carries the signification of aspects of reality, being in line (as opposed to the named 'line'), angularity, curvature, and enclosure.

This last feature is the essential delineator of object as distinct. When the young child scribbles it quickly achieves a mark which is closed upon itself. In so doing it achieves an abstraction from 'reality'. It is the first intellectual exercise of the infant. It cannot 'name' the quality it has discovered but it can perceive it, this ability to perceive without naming is a generator of sensation, which is a body state.

The Idea of separate aspects of a reality in which there are no distinguished aspects prior to the act of depiction precedes Thought.

The Thought is dependent upon the prior act of distinction. This obtaining of an idea which cannot be expressed in language gives rise to emotion.In the obtaining of information there is an attention to the state of the Self. The line stands for an objectness which is "there" and "not there" at the same time.

It is this sensation generating paradox which is the root of the ability of the picture to be 'about' some 'thing'. It is at the root of the ability of the picture to have a subject and a content, the two are not the same thing. The subject is named and the content is experienced.

By the act of drawing the young child perceives an objective self-value, since in the reciprocal transaction values which constitute perception the affordance points two ways and the meaningfulness of the action-Self is perceived long before there can be adequate words to express this quality.

Attention to the linear traces is accompanied by attention to the Self as a value. From self-awareness there is a progression to attention to the Self as a worth for attachment. This Self-reference as worth for attachment, this objectified action-self, is we believe, a more accurate description of that ill-defined concept which plays such a large part in psychodynamic theories, the Ego.

Since there is a difference in perception between the perceived object as an invariant structure of meaningfulness and the graphic trace, attention is constrained to the difference/sameness ratio as meaningful in itself. In the process of choosing the series of relational lines which will invariantly express the desired pictorial object the artist is obliged to pay attention to the object, in terms of variant aspects of angularity, occlusion and so on. The meaningfulness of the environment/artist perceptual synergy then includes attention to aspects of the artist, as much as the artist attends to aspects of the pictorial scene.

The awareness of self-in-action which accompanies all perception constraints the perceiver to an awareness of the meaningfulness of the Self-in-action as distinct and constituted of aspects. The artist is thus constrained into being a self-observer constituted of aspects and no longer an undivided whole.

The drawing is an attempt to find an aesthetic solution to a problem, the nature of the problem being essentially to discover the problem. Aesthetics is a generalizing function, through the process of art (the graphic activity) the patient as artist learns to discover the interactional nature of relations and through the development of an aesthetic function to order these relations toward closure in autonomy. In the interactive objectification of value relations in the act of drawing the interaction itself moves towards its own closure. This process, since it is an interaction between the artist as individual and the drawing which is a representation of the act as an Other leads to the search for structure relations which constrain the artist to self-observer in aesthetic process. In the hands of the trained art-therapist the client is constrained to become their own therapist through the interaction with the drawing as Self-as-Other.

References

Baudrillard, J. (1972). *Pour une critique de l'économie politique du signe*. Paris: Gallimard.

Bohm D., F.D. Peat (1987). *Science, order and creativity*. London: Routlegde & Kegan Paul.

Gibson J.J. (1979). *The ecological approach to visual perception*. Boston: Houghton Mifflin.

Gibson, J.J. (1966). *The senses considered as perceptual systems*. Boston: Houghton Mifflin.

Hegel, G.W.F. (1977). *Phenomenology of spirit*. Oxford: Clarendon Press.

Koestler, A. (1964). *The act of creation*. London: Hutchinson.

Marx, K. (1973). *Grundrisse: foundations of the critique of political economy*. Harmondsworth: Penguin.

Maturana, H.R. (1980). Biology of cognition. In: H.R. Maturana, F.J. Varela, *Autopoiesis and cognition*. Boston: Reidel.

Maturana, H.R. (1988). Reality; The search for objectivity or the quest for a compelling argument. *Irish Journal of Psychology, Special Issue on "Radical Constructivism, Autopoiesis and Psychotherapy"*, V. Kenny (ed.), **9**, 1, 25-82.

7

How to Make Use of Oneself as an Instrument in Systemic Therapy

Max J. van Trommel

Abstract. This chapter deals with a way in which a systemic therapist can make use of himself or herself as a person in making an assessment and also as a tool in making interventions.

Pivotal in the domain of therapy is the encounter between people. How a therapist uses himself or herself in such an encounter is in this chapter highlighted from a theoretical point of view. Concepts of Bateson and Maturana prove to be useful in explaining the process of a therapeutic encounter. Especially their concepts of action, meaning, reflexivity and structural coupling.

Introduction

THIS CHAPTER DEALS with a way in which a systemic therapist can make use of himself or herself as a person in making an assessment and also as a tool in making interventions. Pivotal in the domain of therapy is the encounter between people.

How a therapist uses himself or herself in such an encounter is difficult to describe because it is partly related to one's personality traits. However there are also factors available which can be described as one's expertise. I will try to elaborate on the theoretical aspects of these factors which contribute to the expertise of a therapist and are of use in his or her daily practice.

Differences and Abnormality

In my opinion essential contributions to the theory of understanding a therapeutic encounter have been given by Bateson and Maturana.

The first and most important steps to apply Bateson's ideas in practice has been set by the Milan team. Their concepts of hypothesizing, circularity and neutrality are well known examples (Selvini a.o., 1980, Cecchin 1987). Maturana's ideas about the functioning of living organisms gave further directions to the development of the concept of autonomy.

Bateson (1972, 1979) described a way of understanding the process of human observations. He also described how people, after having made observations, come to processes such as acting and thinking. This proves to be possible by following a repeating pattern of making observations followed by giving meaning to those observations.

According to Bateson it is only possible to make observations by observing a special kind of differences: i.e. differences which are labeled by the observer as being different form a preceding observation. "Differences which make a difference".

In therapy one can make use of this principle. If a therapist during the interview searches for differences, e.g. asks for the different opinions family members may have concerning a certain topic, he or she may learn from the answers that the family members are not able to discover clear differences from each other. If however the therapist expected to get informed about differences between family members he or she might become aware of a relevant difference between the anticipated and the real answers. He or she discovers in that case a difference that makes a difference, which may trigger the hypothesis that the therapist touches a problematic area. Not being able to discover differences does not mean in general that the therapist touches a problematic area.

However when he or she expects to obtain different opinions and the client or clients do not perceive differences it can be a sign of touching a problematic area.

On the other hand it is also possible that the therapist obtains answers to his or her questions which contain only minimal or no relevant differences, while his or her client or clients perceive these differences as crucial. This may also be a sign of a problematic area, because again there is a difference between the anticipated and the real answers.

The following example may clarify this. A couple has been entangled in a symmetrical fight concerning opposite opinions. While listening to the conversations of the couple I heard different opinions on the content level, but on the process level I couldn't discover any difference. Both partners told me that they experienced some improvement in their relationship. The two spouses were convinced that their relationship had improved because of the changes the other spouse had achieved. This was perceived as a big difference in explanation. I however discovered no difference. According to my opinion both partners made the other spouse responsible for the improvement. In such a case where the spouses are not able to share the opinion of the therapist, who sees no difference in their opinion, the therapist may become aware of a relevant difference and may make use of this observation as a possible sign of a problematic area.

By analyzing relevant differences of meaning between the client and the therapist, the therapist has a tool for assessment.

Action and Meaning

Bateson differentiated between making observations and giving meaning to observations. A meaning can be seen as a subjective interpretation which the observer attaches to his or her observation.

Making observations and giving meaning to observations are two different processes which are to be seen as separate from each other. The domain of making observations is different from the domain of giving meaning to observations. This difference has its relevance in practical application.

While being in contact with a client a therapist has the possibility to observe his or her own emotions. Apart from making an observation about emotions the therapist has the possibility to give meaning to this observation. Let me give an example.

A therapist meets a couple which continuously struggles with symmetrical fights. After a while he or she experiences his or her position as unpleasant and hopeless. When one experiences himself or herself in a similar position in relation to other couples with symmetrical fights, one might connect a special meaning to this experience e.g. the meaning that this special experience is specific when meeting people with symmetrical fights.

The unpleasant and hopeless feeling thus became a sign for a specific encounter. The attached meaning together with that special emotion has then become a diagnostic tool in perceiving a symmetrical fight.

Reflexivity between Action and Meaning

Another concept derived from Bateson is the reflexive relationship between making observations and giving meaning to these observations. This means that every action is followed by a meaning and will in its turn be followed by another action and so on. Giving meaning to a certain action can be seen as the start of the following action and vice versa. In this way one can understand that human behavior is guided by the meaning attached to behavior. A meaning linked with an action leads to the next action: this action in its turn guides the accessory meaning and so on.

Action and meaning relate to each other in a reflexive way. In this context action can be considered as e.g. performing a task, making an observation, thinking etc.

Example: When I come home at night and find the living room in a mess, while I prefer my room to be clean and in order I first observe a meaningful difference with the situation I like the best. Besides this I attache to this observation the meaning that I enter my own house. This special meaning will guide my action in such a way that I will start to tidy up the room. Would I enter my friend's room with the same chaos a different meaning of "I am here as a guest" would lead the action in a different way probably by not cleaning the room. This may clarify that a different meaning guides a different action.

On the other hand if I would start to clean my friend's room although I first decided not to interfere, it could result in the meaning "I feel at home in my friend's room". In this case it was the action of cleaning the room which led to a different meaning.

In therapy this means that a therapist should not only direct the attention to problematic behavior but also try to understand the meanings attached to problematic behavior. To understand meanings during an encounter with another person proves to be possible by making use of the concept of structural coupling.

Autonomy and Structural Coupling

Maturana (1975, 1978) and Varela (Maturana & Varela, 1987) applied themselves to the concept of the autonomy of living organisms. Though their observations are related to the functioning of the human being or the human mind, some of their conclusions may be applied to the way people group together and form social systems. To a certain extent their views may be applied to the assessment and treatment of social systems such as couples and families and other therapeutic systems.

Living systems are, according to Maturana and Varela governed from the inside out. Every living system behaves as an organizationally closed unit which so to say extend its tentacles in order to investigate possibilities to interact with the environment.

When living systems are in contact with each other they scan each other during an ongoing process, which has been called by Maturana the process of structural coupling. Crucial in this theory is that living systems are only able to orient themselves towards the environment from a structure determined position. This means that assessment of the environment is only possible by a process of structure determined scanning. An eye e.g. can be seen as a system which because of its receptivity to light impulses, is able to scan the environment for light signals only. This means that a given system can only interact with those features of its environment which are compatible with its structure. An eye is only able to observe light but no sound waves.

In a similar way a therapist is bound to make use of his or her own structure during the process of structural coupling with the client or clients he or she meets during a therapeutic contact. One has however to realize that one is only able to scan the environment according to one's own structure, which means that one can only make observations which are compatible with one's own set of observational instruments.

Assessment and Structural Coupling

Structural coupling can be compared with a dance performed by two people. In this example the dance functions as an opportunity for the dancers to experience how the other person dances and how the two participate in the dance together.

One should realize that the observation of the dance has to be distinguished from the performance of the dance. An observer is able to observe an object such as a dance, while he or she is actively taking part in that dance. Processes can be performed but can not be observed directly.

A process however can be observed through its result or its effect. This can be clarified by an arbitrarily chosen object such as a tea-cup.

A tea-cup can be seen as the end result of a production process. It can be observed by an observer as a result of a production process.

The production process itself of the tea-cup cannot be observed while one is producing the tea-cup; however this process can be observed indirectly, by being aware of differences that occur in the producer/observer him- or herself, like e.g. differences in fatigue between various moments. Then the production process itself has become indirectly the object of observation. In a similar way we can analyze a dance.

A dance can be understood as a production process which is produced by the cooperation of the participants. The result of the performance can only be observed indirectly. Each separate step gives no information about the dance. Only after having linked up the separate steps is it possible to recognize a pattern to which the name of a dance can be attached. Also in this case it is only possible to observe the result of a process as an object indirectly.

Relation between Object and Process

Goudsmit (1989) describes the constitution of an object as intermediate between two processes; according to him an object can be seen on the one hand as the result of naming a process, on the other hand as the prescription of the following process. This means that an object has alternatingly a descriptive and a prescriptive character.

For our present purpose we can, likewise, define a process as something that takes place between two objects. In this way a process can also be understood as having a prescriptive as well as a descriptive character.

The participants of a dance are therefore able to observe their dance by comparing certain objects over a time interval. They can e.g. compare differences in their own mood between the start and a later moment of their dance. This produces the opportunity to get informed about the kind of dance they have developed.

This kind of information will also have a prescriptive function in performing the next part of the dance. In this way a participant of a process is able to reflect upon the process in which he or she participates, which makes it possible that the process has become the object of observation. The process can be observed indirectly by the participants by making observations of previously not existing differences in the participants.

When a participant in a process e.g. is able to observe himself or herself changing, this means that one is able to discover differences which were previously not present. One may then conclude that a process is going on, as well as what kind of process is developing. This means that the differences the observer observes from the inside out informs him or her about the kind of relationship one maintains at that moment with somebody or something else.

When this will be applied to understand a therapeutic encounter we have to be aware of at least two domains in which a therapeutic encounter takes place. In the first domain therapist and client have their conversation. They are active in the domain of the production of a process of conversation. Observing the process of conversation is another domain. During the process of conversation the result of their conversation can be seen as the object which has as well a descriptive as a prescriptive force. The development of the conversation provides the opportunity to define the foregoing part of the conversation and will also influence the next part of the conversation.

However, in order to assess the conversation during the process of conversation the therapist has to interrupt the conversation. This interruption may be very short or longer. During the interruption he or she is able to assess the process of one's own conversation by comparing differences in him or herself which have taken place during the conversation. This process of observation of differences takes place on another level of abstraction than the process of conversation (figure 1).

```
----->P_{i,j}----->O_{i,j+1}------>P_{i,j+2}------>O_{i,j+3}------>P_{i,j+4}------->
                      /\                            /\
                      ||                            ||
----->O_{i+1,j}--->P_{i+1,j+1}---->O_{i+1,j+2}---->P_{i+1,j+3}---->O_{i+1,j+4}---->
```

Figure 1.

In figure 1 time runs from left to right indicated as j, j+1, j+2, etc. In the domain of i the dance O is performed. P stands for the production process of the dance. In the domain of i+1 the process of observation of the performance of the dance has been symbolized. In a therapeutic contact this process of the production of a dance can be described as the process of structural coupling. In the domain of i+1 the dance can be observed as an object in which P symbolizes the process of observation. The vertical arrows symbolize the linkage of the development of the dance with the development of the observation of the dance by its participants.

When the differences between $O_{i+1,j+1}$ and $O_{i+1,j}$ represent internal changes in the participant of the dance, the participant is able by observing the difference between one's present mood and the mood at the start of a therapeutic contact, one gets informed about the dance $O_{i,j+1}$ he or she is performing. This means that when a participant of a dance is able to observe his or her own changes during that dance he or she has an opportunity to learn about the specific kind of dance he or she is dancing.

The foregoing process can also be applied to describe a psychotherapeutic contact. In systemic assessment and therapy one of the therapist's goals is to explore the patterns of ideas, modes of experience, behaviors and relationships of one's clients (Weber et al. 1988). Therapist and clients make use of language to find a consensual domain in which they can exchange ideas. Both parties involved will produce thoughts, secondly name their thoughts and eventually pronounce them which will lead to new thoughts and so on. Both parties will in another domain evaluate their contact, even at an unconscious level, which means that they gather information about the meeting they construct together.

The therapist has to be able to reflect upon this meeting by observing differences between a certain amount of "beacons", as a ship or an airplane may locate itself by noticing the differences in relation to the orienting beacons. By registration of the observed differences one experiences during the therapeutic meeting, one has the opportunity to get informed about the relational pattern one has been developing with one's clients.

Example: if a therapist during his or her contact observes that his or her mood is getting depressed, he or she may deduce from this that it could be possible that the client conveys his or her depression to the therapist. By observing the differences of one's mood between various moments one gets informed about the way in which therapist and his or her clients are relating to each other.

Reflection in Action

During the process of structural coupling the therapist directs his attention first to the performance of the dance and secondly to the assessment of that dance. That means a reflection in action.

If the therapist during the encounter is able to analyze to what extend his or her client or client system influences the relational pattern this will in an impressive manner give information about the way in which the client performs and thus forms patterns of significant ideas. That is to say that a therapist can make use of an analysis of the development of his or her own behavior and sets of meanings during the encounter in order to get informed about the functional and dysfunctional patterns of his or her clients.

A central condition for a functional course of such a process is the therapist's capacity to analyze and direct his or her own part in this interaction. After having been part of the system one must climb to another domain to be able to describe one's experiences during the time one took part in the process. This expertise of moving back and forth during the encounter is one of the core qualities a therapist must have developed as his or her specialization.

If a therapist devotes himself or herself as described above her or she has an opportunity to get very fast to the heart of the matter of his or her clients' problems. In order to realize this one has to register a broad area of signals. One has to be sensitive to verbal, non-verbal and emotional impressions. One should also be able to change deliberately one's behavior in order to get informed about the ways in which one's clients are able to react on these changes. This means that one should be able to be flexible in alternating a neutral with a non-neutral position.

Being neutral means obtaining a stance of curiosity as described by Cecchin (1987) or reflecting in action such as mentioned before; being non-neutral means introducing new behavior or new explanations or providing perturbations during an encounter.

Up till now it has not been stressed that a systemic therapist has to work with one's own emotional experiences during a therapeutic session. On the contrary most authors expressed the need to keep a neutral position without reacting on a non-neutral or emotional level. I agree with those authors who state that a therapist can make assessments about his or her observations only from a neutral position. By getting structurally coupled and by following the dance fluently one finds oneself in a non-neutral position which gives the opportunity to experience the actions, emotions, explanations and narratives of one's clients.

In my opinion a therapist should make use of neutral as well of a non-neutral position. When during the encounter one experiences a change in one's own mood one has to be able to analyze these changes in mood, behavior, emotions, etc. in order to categorize them. After having become aware of these specific changes one has to decide whether or not and if so how one would react to them. These deliberate reactions are actions in the domain of non-neutrality and could be seen as actions brought forth by counter-transference (Kohut 1985, Van Trommel 1987).

Specific in the way of working of a therapist is the continuous alternation of observing and evaluating one's observations, as well as a continuous alternation of a stance of neutrality or curiosity and of non-neutrality.

References

Bateson, G. (1972). *Steps to an ecology of mind.* San Francisco: Chandler.

Bateson, G. (1979). *Mind and nature: A necessary unity.* New York: Dutton.

Cecchin, G. (1987). Hypothesizing, circularity, and neutrality revisited: An invitation to curiosity. *Family Process*, **26**, 405-413.

Goudsmit, A.L., (1989). The black hole of psychotherapy research. Organizational closure in psychotherapeutic processes. In: E. Rosseel, F. Heylighen, F. Demeyere (eds.), *Self-steering and cognition in complex systems.* New York: Gordon & Breach.

Kohut, H. (1985). *The analysis of the self. A systematic approach to the psychoanalytic treatment of narcissistic personality disorders.* New York: International Universities Press.

Maturana, H.R. (1975). The organization of the living: A theory of the living organization. *Int. J. Man-Machine Studies*, **7**, 313-332.

Maturana, H.R. (1978), Biology of language: the epistemology of reality. In: G.A. Miller, E. Lenneberg (eds.), *Psychology and biology of language and thought*. New York: Acad. Press.

Maturana, H.R., F.J. Varela, (1987). *The tree of knowledge*. Boston, Massachusetts: New Science Library.

Selvini Palazzoli, M., L. Boscolo, G. Cecchin, G. Prata, (1980). Hypothesizing-circularity-neutrality: three guidelines for the conductor of the session. *Family Process*, **19**, 3-12.

Trommel, M.J. van (1987). Een therapie met God als co-therapeut. In: R.W. Trijsburg, F. Verhage (eds.). *Onalledaagse geneeskunde*. Assen/Maastricht: Van Gorcum.

Weber, G., F.B. Simon, H. Stierlin, G. Schmidt (1988). Therapy for families manifesting manic-depressive behavior. *Family Process*, **27**, 33-49.

8

On Blindness and Incomprehension

Arno L. Goudsmit

Blindness

VON FOERSTER (1973, 1984) mentions the phenomenon of our retina's blind spot and argues that it is not so much our blindness itself at this spot that is particular but rather our blindness to this very blindness[28]. We do not see that we do not see:

> "... *this localized blindness is not perceived as a dark blotch in our visual field (seeing a dark blotch would imply "seeing"), but this blindness is not perceived at all, that is, neither as something present, nor as something absent: whatever is perceived is perceived "blotch-less"*." (Von Foerster, 1973, p. 36)

The absence of sight goes unnoticed here. The observing person's visual experience is of a continuous space, as Maturana and Varela (1987, p. 17) put it. A sheet of paper is seen without a patch.

[28]Comparable notions on blindness have been discussed by Culler (1983, pp. 272ff.) in the context of literary criticism.

We may say: the unseen patch on the sheet, as well as the discontinuity of the space that occurs due to the local blindness, belong to the phenomenal domain of a meta-observer. Their going unnoticed to the observer himself is due to his organizational closure as an autonomous individual. He brings forth a world by his own means, not by those of the meta-observer.

Let us extend this line of reasoning for a moment. Given the present situation in which an observer has a local blindness, without seeing a discontinuity, then a particular blindness also occurs to the meta-observer: he cannot see the sheet of paper as the observer himself sees it, viz. as a space (of paper) in which the absence of the patch doesn't appear as a discontinuity. The meta-observer's awareness of the patch effectively impedes him to share the perspective of the observer! Though the meta-observer knows that the observer doesn't see the patch, this doesn't make it invisible to the meta-observer.

The meta-observer relates in a comparable way to the observer's local blindness. Here the meta-observer is also incapable of sharing the observer's perspective. Though the meta-observer knows that the observer doesn't see his own local blindness, this doesn't make this blindness invisible to the meta-observer. The meta-observer's awareness of the observer's local blindness again effectively impedes him to regard the observer's visual performance in the way the observer might regard it himself, viz. as a space (of performance) in which the absence of the local blindness doesn't appear as a discontinuity.

Incomprehension

Now a similar kind of closure, involving blindness and incapacity to share perspectives is the case in psychotherapy research. Here the researcher is bound to an incomprehension of some vital aspects of the therapeutic interaction process.

As therapeutic conversations tend to be concerned with qualities of their actual interaction processes, logical levels (content and process) become indistinguishable to the participants of the therapeutic interaction (therapist and patient). 'That which is said' (the object of the conversation) merges with 'that which is done' (the process of the conversation). Whereas before this merge[29] occurred, a content of discussion was not identified with the process of discussing it, now, at a certain moment, the two become indistinguishable. This merge is not noticed by the participants. They do not distinguish their lack of distinction. Like in blindness, the organizational closure that is occurring here by definition evades the attention of those involved. Once taking place, critical distance is no longer possible. The closure, as performed by those involved in it, exists as a phenomenon only for those external to it, like the blindness to blindness only existed for the meta-observer.

Like the meta-observer, however, the therapy researcher also obtains a particular incomprehension, due to the occurrence of the merge: he cannot understand the process and the object of the therapeutic conversation as the participants themselves understand it, viz. as a space (of conversation) in which the absence of the process-object distinction does *not* appear as a discontinuity. Neither is the researcher capable of sharing his perspective with the therapy participants. Though the researcher knows that they do not make the process-object distinction, this doesn't make the process and the object indistinguishable to him.

How, then, is the researcher to share the perspective of the therapy participants? It is comparable to a meta-observer's attempts to see what the observer sees. The meta-observer may direct his eyes and his look at the sheet of paper in such a way, so as to create for himself the same local blindness as is the case for the observer. Likewise, the therapy researcher may construct for himself the same absence of process-object distinction, as is the case for the therapy participants. The researcher may share their perspective by creating a self-referential conversation in which he participates. This can be constructed by entering into a conversation with the participants, e.g. by means of a separate encounter with the therapist. He can start to exchange perspectives with the therapist about his therapeutic conversation with the client. Such a

[29]Notice that the present merge between object (content) and process is similar to the object-method merge of our first chapter.

conversation entails a quest for consensus by therapist and researcher about their perspectives.

Since both see different things and are blind to different things, such a quest can be directed into an exploration of their present process of seeking consensus. In other words: when this new conversation also becomes self-referential it will yield a new merge between process and object. In particular, it will be concerned with exploring the difficulties for both parties to share their respective brands of blindness. Then the new conversation will be concerned with itself in a way that highly resembles the self-referentiality of the therapeutic conversation that was the original topic of discussion. This resemblance is typical for all varieties of transference. Precisely by this phenomenon it is possible to orient one's conversation partner effectively towards the things and issues that evade being put into words. It is what also happens in the so called 'parallel processes' of supervision sessions. Supervision, therefore, may turn out to be our most valuable research tool. The local blindness, the incomprehension of the other person's point of view, is here explored as a source of knowledge.

The logic of psychotherapy is one of blindness and incomprehension, and the best thing we can do is to take them serious as phenomena in their own right.

References

Culler, J. (1983). *On deconstruction. Theory and criticism after structuralism*. London: Routledge & Kegan Paul.

Foerster, H. von (1973). On constructing a reality. In: W.F.E. Preiser (ed.) (1973). *Environmental design research, vol 2.* Strousbury: Dowden, Hutchinson & Rose.

Foerster, H. von (1984). Principles of self-organization - in a socio-managerial context. In: H. Ulrich, G.J.B. Probst (eds.), *Self-organization and management of social systems; insights, promises, doubts, and questions*. Berlin: Springer.

Maturana, H.R., F.J. Varela (1987). *The tree of knowledge. The biological roots of human understanding.* Boston/London: Shambhala.

About the Contributors

Arno Goudsmit works as a psychotherapy researcher at the University of Groningen and at the Center for Systems Research, University of Alberta, Edmonton, Canada. He studied psychology and philosophy, and he has published on psychotherapy research and cybernetics. Address: Dept. of Foundations and History of Psychology, University of Groningen, Oude Boteringestraat 34, 9712 GK Groningen, The Netherlands.

Kim James has degrees in cybernetics, neuropsychology and fine art. He has a 30 years of experience as a consultant and trainer. He is director of the management consultancy institute P.S.I. International, and he is consultant director for a long-term training programme at the French National Training Institute (INFIPP) in Dijon. Address: P.S.I. International, 23 Hickmire, Wollaston, Northampton, NN97SL United Kingdom.

Vincent Kenny is director of the Institute of Constructivist Psychology in Ireland and director of Psychotherapy Training at the Department of Psychiatry, University College, Dublin. He is an editor of the International Journal of Personal construct Psychology, and has published widely on psychotherapy and epistemology. He studied philosophy. Address: Institute of Constructivist Psychology, 11 Burgh Quay, Dublin 2, Ireland.

Gerhard Portele was subsequently teacher, studied in Mannheim and Heidelberg sociology, psychology and philosophy and was research assistant at Mannheim University. Since 1977 he is professor of Educational Sciences at Hamburg University. Since more than ten years he also works as a gestalt therapist and as a teacher of gestalt therapy. Address: Universität Hamburg, Interdisziplinäres Zentrum für Hochschuldidaktik, Sedanstrasse 19, 2 Hamburg 13, Federal Republic of Germany.

Henri Schneider is a research associate at the department of Clinical Psychology at the University of Zürich, Switzerland. Starting from Piaget's theory, he became interested in models of change and evolution in physics and biology. The primary emphasis in his work is on the use of models from these domains as heuristics for "discovering" pathways of change in psychotherapy. With Urs

Wüthrich (University of Berne) he is working out a procedure for the systematic investigation of change processes in clinical material. Address: Universität Zürich, Abteilung Klinische Psychologie, Schmelzbergstr. 40, CH-8044 Zürich, Switzerland.

Max van Trommel is psychiatrist. Before graduation as a psychiatrist he worked as a general practitioner and as an associate professor in general medicine. He is head of a department of psychotherapy in a mental health center. He also leads a teaching department in systemic therapy. He has given workshops and presentations in several countries and published about general medicine, systemic therapy and on teaching. Address: Regionaal Centrum voor de Geestelijke Gezondheidszorg R.N.O., Schiekade 121, 3033 BK Rotterdam, The Netherlands.

Name Index

Subject Index